INTRODUCTION TO ULTRAHIGH ENERGY COSMIC RAY PHYSICS

Frontiers in Physics

For more information about this series, please visit: [www.crcpress.com/Frontiers-in-Physics/book-series/FRONTIERSPHYS]

Introduction to Ultrahigh Energy Cosmic Ray Physics

Second Edition

Pierre Sokolsky
and
Gordon Thomson

CRC Press
Taylor & Francis Group
Boca Raton London New York

CRC Press is an imprint of the
Taylor & Francis Group, an **informa** business

Second edition published 2020
by CRC Press
6000 Broken Sound Parkway NW, Suite 300, Boca Raton, FL 33487-2742

and by CRC Press
2 Park Square, Milton Park, Abingdon, Oxon, OX14 4RN

© 2021 Taylor & Francis Group, LLC

First edition published by Westview Press 2004

CRC Press is an imprint of Taylor & Francis Group, LLC

ISBN: 978-0-367-17384-5 (hbk)
ISBN: 978-0-367-15117-1 (pbk)
ISBN: 978-0-429-05515-7 (ebk)

Typeset in Adobe Caslon Pro
by SPi Global, India

Contents

List of Figures

Acknowledgments

The authors would like to thank their collaborators in the Fly's Eye, High Resolution Fly's Eye, and Telescope Array experiments for many years of effort in the study of ultrahigh-energy cosmic rays, in particular their co-spokespersons Masaki Fukushima and Hiroyuki Sagawa, and their colleague Charlie Jui. Their graduate students and postdocs deserve particular thanks. The material presented in this book is the result of all of their work. Without the support of the U.S. National Science Foundation, none of this would have been possible. The University of Utah and Rutgers University deserve particular thanks. Pierre Sokolsky would like to thank the Santa Cruz Institute for Particle Physics and the University of California, Santa Cruz Physics Department for providing access to their facilities during a sabbatical leave.

Gordon Thompson would like to dedicate his part of this book to his wife, Tricia Thomson, and thank her for her support during the writing and for her help in editing his chapters.

The first edition of this work, and hence this new edition, would not have been possible without the support and encouragement of Tom O'Halloran and Clicerio Avilez, and particularly Susan Sokolsky's indispensable and detailed editing. Finally, Pierre Sokolsky would like to dedicate his contribution to the memory of Rich Orr, whose friendship and infectious enthusiasm for science made all the difference.

List of Abbreviations

ADC	Analog to Digital Conversion
AERA	Auger Engineering Radio Array
AGN	active galactic nuclei
ANITA	Antarctic Impulsive Transient Antenna
ARA	Askaryan Radio Array
ARIANNA	Antarctic Ross Ice Shelf ANtenna Neutrino Array
BSD	boron doped scintillator
CMBR	cosmic microwave background radiation
DC	Direct Current
EAS	extensive air shower
EM	electromagnetic
FADC	flash analog to digital converter
FD	fluorescence detector
GRB	gamma-ray burst
HiRes	High Resolution Fly's Eye
ISS	International Space Station
Kpc	kiloparsec
LHC	Large Hadron Collider
mb	millibarn
Mpc	megaparsec
mwe	meters water equivalent
PAO	Pierre Auger Observatory

PMT	photomultiplier tube
RHIC	relativistic heavy ion collider
RICH	ring-imaging Cherenkov
SCD	silicon charge detector
SD	surface detector
S/N	signal to noise ratio
TA	Telescope Array experiment
TALE	Telescope Array Low Energy Extension
TCD	timing charge detector
TRD	transition radiation detector
UHECR	ultrahigh energy cosmic rays
VEM	vertical equivalent muon

Introduction to the Second Edition

More than 30 years have elapsed since the first publication of this book, in the Addison-Wesley Frontiers of Science series. A second printing, by the Westview Press in 2004, was a reprint, the introductory nature of the material being thought of as still useful to the student or newcomer to the field. However, much has changed since then and this new edition, while still keeping the focus on an introductory and pedagogical approach, has been rewritten and updated to reflect the wealth of new technical developments and scientific results. The authors of this edition have had a long history of scientific collaboration on the High Resolution Fly's Eye (HiRes) and Telescope Array (TA) projects, which allows us to broaden and deepen the discussion of the subject matter.

We follow the general structure and outline of the original text but include several new chapters and new material, and update the discussion while keeping to the original philosophy of the first edition. The primary emphasis is on a straightforward pedagogical approach with illustrative examples from the past and current experimental situation. Where appropriate, older, now classic, experiments are used to introduce the student to the issues at hand. As in the first edition, there is no pretense of completeness. Many of the examples are drawn from the authors' experience with the HiRes and TA projects, and there is inevitably a bias towards using these for illustration. There are many excellent scientific reviews of the field that the interested student can consult to obtain a more balanced view of the current status. We have also kept many of the original references, as we feel that not only is it useful for the student to be able to see how the subject has developed, but also that, in many cases, the early references are more pedagogically clear since they treat what was, then, a novel result.

Some areas, such as VHE gamma-ray astronomy and neutrino astronomy, which were originally in the text, are now important fields in their own right to which we can no longer do justice in such an introduction. Again, the student is directed to many excellent reviews and texts that are now available.

That being said, the book remains addressed to students and researchers in astrophysics, high energy physics, cosmic rays, or astronomy who wish to acquire an introductory understanding of the current issues and experimental techniques in the field of ultrahigh energy cosmic rays.

1

SURVEY OF ULTRAHIGH ENERGY COSMIC RAYS

1.1 Introduction

For our purposes, we define cosmic rays to be charged subatomic particles originating beyond the Earth's atmosphere. This book considers questions relating to cosmic rays at the highest observable energies, where the particle flux is so low that indirect detection methods must be used; that is, energies >0.01 EeV (10^{16} eV). Such cosmic rays have been observed up to ~3 × 10^{20} eV or 50 J. This is a macroscopic energy (a proton of this energy would have a relativistic mass equal to the mass of a paramecium!). The very existence of particles with such energies is a mystery and leads to the question of the nature of the accelerator.

There are many problems in the study of ultrahigh energy cosmic rays (UHECR), but there are two central questions. The first is a result of the fact that cosmic rays on the Earth's surface have been observed from MeV energies up to energies of 100 EeV; that is, the enormous range of energies of this phenomenon. Sources in either our galaxy or other galaxies produce particles with energies ranging over 14 orders of magnitude. Acceleration mechanisms for higher and higher energies become more and more difficult to envisage. The first central question is thus that of origin, which in turn implies an understanding of the acceleration mechanisms involved. That subject is the main concern of this book. The second central question relates to the fact that the energy density of cosmic rays is a significant fraction of the energy density of the universe as a whole. As we shall see, it is of the same order as the energy density of starlight and the galactic magnetic field. It follows that cosmic rays must play an important role in the overall energy balance of the universe. Here, we will largely sidestep this "ecological" question, even though it is also of great interest.

The study of the sources of UHECR reduces to understanding the three different kinds of information that we have about them.

- The first kind of information is the nature of the particle and its energy spectrum as observed on the surface of the Earth; that is, the chemical composition of the cosmic rays and their energy distribution.

- The second kind of information is how the cosmic rays propagate through interstellar and intergalactic space to reach the Earth. In their passage through intergalactic space, we know that they traverse magnetic fields that affect their direction, and that they interact with dust and gas particles and photons, ranging from the primordial black-body radiation to starlight. These interactions cause energy losses as well as the breakup (spallation) of any heavy nuclei that may be constituents of the cosmic rays. For the highest energies at which they have been detected, cosmic rays will interact inelastically with the 2.7K black-body radiation left over from the decoupling era in the early expansion of the universe. This interaction will modify both the observed charged particle spectrum and the composition of the primary cosmic ray spectrum, as well as producing secondary lower energy gamma rays and neutrinos. This interaction also implies that, at the highest energies, the sources of cosmic rays must lie within ~50–100 Mpc from the Earth, or the distance to local superclusters of galaxies.

- The third kind of information we can examine is the isotropy, or lack thereof, of cosmic rays with respect to potential sources within the galaxy or exterior to it. At the highest energies, the effect of magnetic fields on particle trajectories weakens and one may expect small to intermediate-scale anisotropies around the actual sources of cosmic rays.

  On a larger scale, one may expect cosmic rays to follow the distribution of matter in the nearby universe, blurred and modified by magnetic sheets and filaments associated with the large-scale structure of the local universe.

As we shall see, answers to questions relating to chemical composition and spectra, propagation, and isotropy all affect each other. This is what makes the study of cosmic ray sources so difficult. What is needed is experimental data that gives answers about the correlations of the three kinds of issue that have just been described.

1.2 Present Knowledge of Cosmic Rays

Before turning to these questions in greater detail, we will briefly survey what is known about cosmic rays and give order of magnitude numbers to orient the novice. The cosmic ray flux falls in many orders of magnitude (see Figure 1.1), from the MeV range to the 100 EeV range. Experimental techniques that are used to detect this flux in different energy ranges are therefore very different. At low energies of less than about 10^{14} eV (100 TeV), the flux is sufficiently high for direct measurements to be performed using space-based spectrometers, calorimeters, and other similar techniques. Recently, these direct measurements have been extended to about 10^{15} eV (1 PeV).

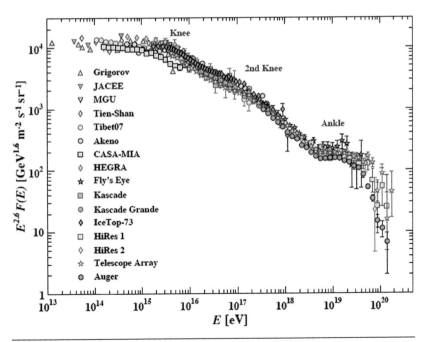

Figure 1.1   Differential cosmic ray flux as measured. Note the rapidly falling flux is multiplied by a power of the energy to bring out the structures. From M. Tanabashi et al., (Particle Data Group), *Phys. Rev. D*, 98, 2018, p. 030001, with permission.

1.2.1 Chemical Composition

The detailed relative abundance of galactic cosmic ray matter at low energies is shown in Figure 1.2. We can compare this composition with the average composition of stellar material in the solar system. Overall, the compositions are quite similar; however, there are some very important detailed differences.[2] Cosmic rays are over-abundant in lithium, beryllium, and boron. The iron concentration agrees quite well with solar system composition, but there is an excess of elements just lighter than iron. There is also an under-abundance of hydrogen and helium. One way to understand these detailed differences is to assume that cosmic rays have the same composition as solar matter at their origin. As they pass through interstellar space, they interact with gas and dust particles, and the heavier nuclei spallate into lighter nuclei. Detailed models show that the abundance of lighter elements in cosmic rays in relation to the composition of solar matter is in agreement with this kind of spallation due to propagation effects. This spallation process introduces a differentiation between two kinds of cosmic ray nuclei: primary (originating at the source) and secondary (produced by propagation effects).

At a more detailed level, we should note that there is some uncertainty about the relative solar abundances. These are different in detail if measured by the spectral lines in the solar photosphere or in chondritic meteorites, which are thought to better represent the initial solar

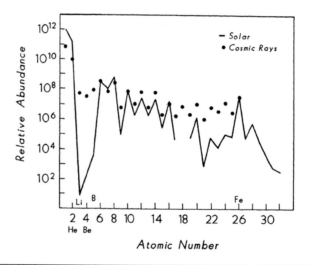

Figure 1.2   Relative abundances of solar and cosmic ray material.

system abundances. Primary cosmic rays not originating in the solar system could, of course, also have an altered composition depending, for example, on the kind of supernova from which they originated.[3]

At higher energies (TeV to PeV), the composition of cosmic rays is approximately 50% protons, 25% α particles, 13% CNO, and 13% Fe.[4] The relative proportions will change slowly with energy as the individual elemental spectra have slightly different shapes (see Figure 1.3).

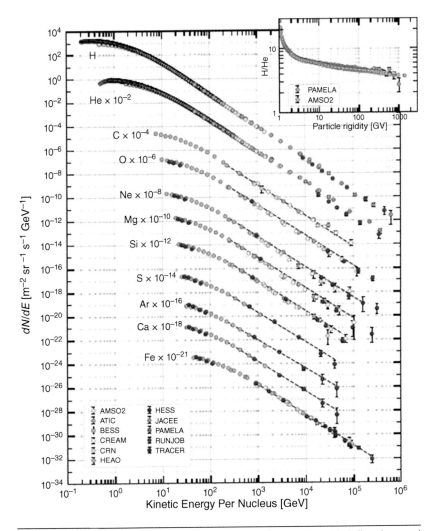

Figure 1.3 Individual fluxes of cosmic rays at lower energies as measured by balloon-borne and ISS based experiments. From Tanabashi et al., (Particle Data Group), *Phys. Rev. D*, 98, 2018, p. 030001, with permission.

Electrons comprise roughly 0.005 of the proton flux and gammas on the order of 0.001 of the cosmic ray flux at these energies.

1.2.2 Spectrum

The overall spectrum of primary cosmic rays obeys an approximate power law with a break and a change of the slope at around a PeV (see Figure 1.1). The fact that such an energy dependence exists over many decades is important in restricting possible acceleration mechanisms, since the source of the cosmic rays must be such as to generate a power law spectrum. This was important in the early theoretical attempts to narrow down the possible sources. Note, however, that the individual nuclear species are now known to have a more complex energy dependence with multiple breaks in the power law spectrum whose nature is still unclear, and the earlier models now appear to be too simplistic.

The flux of primary cosmic rays falls from approximately 1 particle/m^2-s-MeV at the lowest energies to 1 particle/km^2-century at the highest energies. This magnitude of flux implies that the energy density in ultrahigh energy cosmic rays is very large. If the energy density that we observe on Earth is similar to that existing in extragalactic space, a significant component of the total energy of the universe is in cosmic rays. The cosmic ray energy density integrated over all energies turns out to be approximately 1 eV/cm^3. For comparison, starlight has an energy density of 0.6 eV/cm^3 and the energy density of the galactic magnetic field is 0.2 eV/cm^3. Cosmic rays form a major constituent of the interstellar medium. At the highest observed cosmic ray energy, the resultant energy density causes great problems in terms of source energy. The flux at 100 EeV implies an energy density of 10^{-8} eV/cm^3. If we also assume that these cosmic rays fill the local supercluster of galaxies and have a lifetime of about 10^8 years, then the sources of these cosmic rays in the supercluster must pump out approximately 5×10^{41} eV/s at 100 EeV to keep the flux constant.[5] This required energy input is comparable to the entire radio band energy output of the galaxies M87 or Cen A. It is clear that cosmic ray sources cannot follow a black-body radiation spectrum! There must be nonthermal mechanisms for accelerating particles to these enormous energies.

1.3 Candidate Sources

The presently accepted view is that ultrahigh energy cosmic rays are created and accelerated in active cosmic objects. These include: supernovae, pulsars, active galactic nuclei (AGNs), quasars, star-forming galaxies and gamma-ray bursters (GRBs). We will briefly describe the distance and energy scales involved. Our galaxy has a radius of approximately 10 kiloparsecs (kpc) (1 parsec is equal to 3.26 light-years) with the sun located approximately that distance from the galactic center. The thickness of the galactic disk is approximately 100 parsecs. Possible energetic sources of cosmic rays in the galaxy include: supernovae explosions, pulsars, and the galactic nucleus, which we now know contain a supermassive black hole. Our galaxy is a member of the local cluster of galaxies, which has a scale of approximately 2 megaparsecs (Mpc). The local cluster is, in turn, a part of the local supercluster with a scale of 30–50 Mpc centered on Virgo, approximately 20 Mpc away. The local supercluster has a large number of highly energetic radio galaxies that can certainly produce the required energy output for lower energy cosmic rays. Beyond the local supercluster are other superclusters extending to the edge of the visible universe. Clusters and superclusters are connected to each other by filamentary threads of gas and magnetic fields, forming the so-called "cosmic web." The edge of the universe is defined by the distance at which the velocity of recession of galaxies is equal to the speed of light. This corresponds to a distance of approximately 5,000 Mpc, assuming a Hubble constant of 60 km/s-Mpc.

There exist possible sources of high energy particles inside the local supercluster and beyond that are not present in our own galaxy. These include radio galaxies, star-forming galaxies, AGNs, and quasars. All of these are characterized by highly energetic galactic nuclei that liberate vast amounts of energy in the form of electromagnetic radiation and, presumably, high energy particles. Many of these radio galaxies have extended emission hot spots that occur beyond the main body of the galaxy itself, so called "jets." A classic representation of such a jet is M87. The nucleus of such a galaxy is believed to contain the basic energy source: a supermassive black hole surrounded by an accretion disk. In the process of conversion of the angular momentum of the in-falling accretion disk, one or more jets approximately perpendicular to the disk are formed. In ways not completely understood, particles can be accelerated by electromagnetic processes in the jets and ejected into extragalactic space. This is demonstrated by the existence of extended radio sources (in particular, double radio sources) whose radio hot spots are clearly due to synchrotron radiation produced by high energy particles ejected from the main body of the galaxy. Some of these ejected particles may reach our galaxy and be observed by us on Earth. This is particularly the case when the jet axis is pointed towards the Earth (the so-called "BL Lac galaxies"). Energy estimates for the various energy sources over their lifetimes are 10^{49}–10^{51} eV for supernovae, 10^{56}–10^{57} eV for the galactic nuclei themselves, up to 10^{62} eV for radio galaxies, and 10^{58} eV for quasars.[6]

GRBs are some of the most energetic events ever observed and have naturally been suggested as sources of UHECR. These events are believed to be produced by either superluminous supernovae, or the merger of binary neutron stars. GRBs produce copious electromagnetic radiation. GRB 080319B, for example, was an event with electromagnetic energy approximately equal to the rest mass of the Sun, if the energy were isotropically radiated. Evidence indicates that the radiation is emitted in jets, however. An interesting model for GRBs is that of a collapsar, where an extremely massive star collapses to a black hole and, in the process, produces two relativistic jets. Ultrahigh energy cosmic rays could be accelerated by the shock waves produced by the core collapse, or in the jets themselves. However, due to the intense photon field surrounding such a GRB, photonuclear reactions will produce significant energy losses for the accelerated hadrons, as well as generating a neutrino flux. Recent limits on the neutrino flux from GRBs have made these scenarios more doubtful.

Neutron stars and black holes with accretion disks (which exist in our own galaxy and in others) are almost certainly candidate sources of PeV energy galactic cosmic rays (together with supernova explosions). It is very difficult to see how they can contribute to the highest energy flux, however.

1.4 Acceleration Models

In general, the basic physics models thought to be responsible for acceleration of cosmic rays fall into two classes: statistical acceleration or direct acceleration. In the case of statistical acceleration, the final energy is gained slowly over many decades of energy. The prototype of this kind of acceleration is the Fermi acceleration model.[7] Fermi originally proposed that particles are accelerated by collision with magnetic clouds in the galaxy (second-order Fermi acceleration); this picture can also be extended to acceleration by shock waves from supernovae, as well as in galactic nuclei, and radio galaxy hot spots and associated jets (first-order Fermi acceleration). An advantage of statistical acceleration models is that the observed power law energy spectrum is achieved in a natural way. In the second-order case, the acceleration is slow and occurs over an extended region of space. In the first-order case, it is generally difficult for energy gained to keep up with energy lost, and we shall see that this restricts possible regions of space where such acceleration can occur to a very small number of candidates.

Direct acceleration, on the other hand, assumes the existence of a strong electromagnetic field. The acceleration is fast, and is particularly applicable to systems such as pulsars with strong rotating magnetic fields producing an induced electromotive force. How a power law spectrum results from such an acceleration mechanism is not obvious. Energy losses may also be a serious problem if the accelerating region is in an area of very high radiation density.

In addition to these astrophysical models, there have been proposals that the highest energy cosmic rays are the decay products of exotic objects such as monopole–antimonopole atoms (monopolonium), or superconducting cosmic strings. Such models also produce a low energy gamma ray background. These proposals have been largely eliminated by current gamma ray flux data.

A major problem with all acceleration models is the energy loss experienced by particles accelerated in dense regions of space, particularly those near supernovae, and galactic nuclei and their jets. There is believed to be a high concentration of optical photons in these regions, on the order of $10^{14}/cm^3$ with temperatures between 10^4K and 10^5K.[8] The region of acceleration can be on the order of 3×10^{16} cm. If cosmic ray photonuclear interaction cross sections are of the order of 3×10^{-31} cm^2, then a significant energy loss will occur in the acceleration region. Energy loss mechanisms include: meson photoproduction (cross section on the order of 10^{-28} cm^2); photonuclear fission processes which break up heavy nuclei (cross section of 10^{-26} cm^2); and electron–positron production. As a result, significant changes to both the spectral shape and the composition may occur for particles produced in such regions, leading to differences between the "injection spectrum" and the spectrum as observed just outside the photosphere of the accelerator. In fact, due to the close coupling between acceleration and energy loss, it may make more sense to define the "injection spectrum" as the spectrum just external to the source which is then modified by propagation effects through the intergalactic and interstellar medium on its way to the Earth.

1.5 Propagation

Assuming that the accelerated particles can escape the source region, how do they get to us? If the particle is produced within our galaxy, it must traverse the interstellar medium to arrive at the Earth. If it is of extragalactic origin, it must traverse the interstellar medium of the galaxy in which it was created; cross the intergalactic medium into the interstellar medium of our galaxy; and, finally, reach the Earth. It is also likely that the space between galaxies in a cluster has different properties, such as an enhanced magnetic field, than the average extragalactic space.

We first consider the interstellar medium. It is composed of clouds of neutral and ionized gases, predominantly hydrogen. The gas is ionized by starlight. Other components include the chaotic and regular galactic magnetic field, starlight photons, and the 2.7K black-body radiation composed of photons left over from the Big Bang.

Galactic magnetic fields are studied by observing the Faraday rotation of the plane of polarization of linearly polarized light. This includes the polarization of radio synchrotron emission and starlight. The nature of the fields is very uncertain beyond the galactic plane. The regular galactic field has a strength of 1 to 3 microgauss; it lies in the galactic plane and is directed toward galactic latitude 90°. The chaotic fields are produced in magnetic clouds generated due to the streaming motion of ionized gas. The magnitude of the chaotic fields is of the same order as that of the regular galactic field.

Cosmic ray particle trajectories are bent and scattered by the regular and chaotic fields, and this produces a diffusive motion of cosmic rays in the galaxy. The characteristic length for magnetic deflection is given by the Larmor radius, as shown in Equation (1.1):

$$R_L = E/300(HZ) \qquad (1.1)$$

where R_L is in centimeters, E is in eV, H is in gauss, and Z is the particle charge. A particle passing through a magnetic cloud of length l will experience an angular deflection of order l/R_L radians.[9] For a cloud with l = 0.1 parsec, H = 1 microgauss and E = 1 PeV, l/R_L is large (order of 1). Cosmic rays of PeV and lower energy can, therefore, become entangled in magnetic clouds and suffer large deflections. These particles will then diffuse through the galaxy with a mean free path equal to the mean distance between magnetic clouds. At higher energies, angular deflections are smaller. Even so, a number of such scatters can produce a large deflection. For N independent collisions, the total angular deflection will become large when $N^{1/2}l/R_L \sim 1$; that is, N is proportional to $(R_L/l)^2$. Since R_L is proportional to E, it follows that the number of collisions needed for a large angular deflection increases with energy like E^2. This implies that the mean free path is also proportional to E^2 and the motion is not diffusive. As a general rule, the effect of chaotic magnetic fields is small at EeV energies and above.

At PeV energies diffusion is important. This means that the time during which the cosmic ray remains in the galaxy (its age) is greater than would be expected for nondiffusive propagation. Diffusion theory gives the relative increase as shown in Equation (1.2):

$$t_D/t \sim 1/2(r/\lambda) \qquad (1.2)$$

where r is the distance from a source to the edge of the diffusion region, and λ is the mean free path. The total amount of cosmic gas traversed by a cosmic ray is thus proportional to r/λ. If a particle originates at the galactic center and diffuses out to the Earth ($r = 10^{22}$ cm) with a mean free path for collision with magnetic clouds of 3 parsecs, through a gas of mean density 10^{-24} g/cm^3, it traverses ≤ 10 g/cm^2 of material. Since the interaction length for protons at these energies is ~ 70 g/cm^2, the proton spectrum will not be affected by diffusion.

The mean free path for spallation of heavy nuclei is much smaller, however, on the order of 10 g/cm². This leads us to expect that the chemical composition of the cosmic ray flux will be altered by this diffusive motion. In fact, one can work backward from the measured composition to the solar system mass composition to obtain an estimate of the diffusion time. At the highest energies, the mean free path becomes proportional to E^2, which implies that r/λ decreases and the diffusion time approaches the nondiffusive time. Fragmentation, therefore, is not such a problem. We can thus expect that "primary" cosmic ray species (i.e., those produced directly by nucleosynthesis) and "secondary" species (such as Li, Be, and B, which are produced by spallation) will have very different spectral energy dependences. This is, in fact, observed.

Few electrons appear in the cosmic ray spectrum as they are effectively absorbed by Compton scattering on optical photons and by the bremsstrahlung radiation. The dE/dx loss for bremsstrahlung is proportional to E^2; therefore, we expect very few high energy electrons, contributing on the order of 1% of the total cosmic ray flux. However, this energy loss process does generate a low energy photon continuum, which can be observed to yield clues to the electron spectrum at the source. At very high energies (VHE) (~10^{15} eV) γ-rays are also absorbed by gamma–gamma interactions on optical photons from starlight and on the 2.7K black-body radiation. The universe is thus opaque to VHE photons beyond approximately a few Mpc.

The intergalactic medium is believed to be permeated by a magnetic field that is on the order of 10^{-2} of the galactic field, though there is little certainty about this. In this case, protons from the center of the Virgo supercluster, for example, will not be significantly deflected for energies greater than 30 EeV. We can thus expect that anisotropy can exist from relatively nearby extragalactic sources at the highest energies for a light composition. For greater distances, directional information will again be lost. A complication also arises due to the observation that fields in galaxy clusters can form sheets and filaments that can guide cosmic rays and potentially form anisotropic patterns that are not directly correlated with matter density. In other words, the large-scale structure of galaxies in the nearby universe can be an insufficient predictor of cosmic ray origin through anisotropy, though it must be a reasonable guide. One needs to add a knowledge of the filamentary structure of magnetic fields.

The detailed structure of the galactic magnetic field is also very uncertain. Recent observations indicate the presence of a galactic magnetic field halo along a perpendicular to the galactic center. Significant local variations up to an order of magnitude above the average field have also been suggested by observations. Even in the absence of significant bending by extragalactic fields, we can expect trajectory deviations of the order of $10°$ due to these galactic field effects.

The estimated lifetime of cosmic rays in the local supercluster is on the order of 10^{10} years. Assuming a matter density of 10^{-29} g/cm^3 in intergalactic space, the average particle will traverse ~0.2 g/cm^2 of material. This implies negligible probability of interaction. At the highest energies, however, there is an important interaction with the relic radiation. The 2.7K black-body radiation has a density of approximately 500 photons/cm^3. Protons with energies on the order of 50 EeV will collide inelastically with these photons. In the rest frame of a proton of this energy, the 10^{-4} eV γ-ray appears to have an energy on the order of 300 MeV, which corresponds to the threshold for pion photoproduction, as was first pointed out by Greisen and, independently, by Zatsepin and Kuzmin.[10] Due to this "turn on" of inelastic interactions at about 50 EeV, one would expect a change in slope or cutoff of the cosmic ray flux at this energy. One expects to see a change in the spectrum at about 30 EeV if particles have traversed the 2.7K black-body radiation for greater than or equal to 10^9 years. The Greisen–Zatsepin cutoff shape, therefore, gives important information on the distance of cosmic ray sources, as well as verifying the universality of the microwave background radiation.[11] If the extragalactic cosmic rays are nuclei, they will also interact with the 2.7K black-body radiation through photo-spallation, producing secondary nuclei. Light nuclei such as He or N have different spallation interaction lengths and thus cut-offs at different (lower) energies. Fe is the most stable nucleus and has a cut-off at approximately the same energy as protons. This effect implies that the universe is opaque to charged particles with energies near 100 EeV for distances greater than 50–100 Mpc.

An important consequence of photoproduction off the 2.7K black-body radiation is the consequent production of pions and muons, which decay into neutrinos and gamma rays.[12] Observation of a cut-off in the charged particle cosmic ray spectrum can then be correlated with observation of a neutrino and gamma ray flux the magnitude and energy spectrum of which is related to the charged spectrum shape and cutoff energy. Note, however, that only neutrinos have the ability to reach us from anywhere in the universe. Hence this "cosmogenic" neutrino flux represents Greisen–Zatsepin–Kuzmin interactions everywhere in the universe, while the protonic flux seen at Earth has a source horizon of ~100 Mpc.

1.6 Direct Measurement Techniques

Below about 0.1 PeV, the cosmic ray flux is sufficiently high to allow direct measurements to be performed on the primary particle. Techniques are similar to those used in high energy physics experiments. Calorimeters, emulsion stacks, tracking detectors, and transition radiation detectors are flown in balloon flights high in the atmosphere, or on satellites and space shuttle flights. More recently, such experiments have been incorporated as part of the International Space Station (ISS). These techniques now yield direct measurements of the cosmic ray flux and composition up to energies of 10^{15} eV. This energy region overlaps accelerator physics, and measurements are yielding clear information on the cosmic ray composition. There are now many major long-duration balloon experiments such as CREAM and the AMS, ISS-CREAM and CALET experiments are collecting data on the ISS with lifetimes of many years. However, due to the decreasing flux, measurements above ~PeV energies will remain indirect for some time to come.

1.7 Indirect Detection through Extensive Air Showers

Above 0.1 EeV, the flux is so low ($<10^{-14}$ per cm²-second-steradian), that direct detection is impossible. The only available technique is to detect atmospheric extensive air showers (EAS) produced by the primary cosmic ray particle. Protons or heavier nuclei interact with the nitrogen and oxygen nuclei in the atmosphere, and generate an EAS composed primarily of electrons, muons, and photons. Instantaneously, it can be pictured as a pancake of particles ~100 m wide and approximately a few meters in thickness moving through the atmosphere at close to the speed of light for many kilometers until it hits the ground. Integrated over its development in the atmosphere, this air shower is composed of charged and neutral particles extending hundreds of meters in width and 10–15 km in length. The number of particles as a function of atmospheric depth in g/cm^3 follows the well-known calorimeter energy deposition curve familiar from high energy physics.

The gross properties of EAS are determined by the electromagnetic interactions; however, some properties depend on hadronic physics and are of interest to particle physicists. The point at which the incident cosmic ray interacts with a nitrogen or oxygen nucleus is determined by its interaction length which, in turn, is determined by the inelastic proton–air cross section. At laboratory energies of interest, this corresponds to the cross section at center of mass energy ≥ 10 TeV (i.e., Large Hadron Collider energies or greater). Once the particle has interacted, the depth at which the EAS reaches a maximum in the number of electrons is also dependent on the multiplicity and inelasticity of the interaction. Therefore, the depth in the atmosphere of X_{max}, the EAS maximum, is sensitive to the details of hadronic interactions. One can imagine using this sensitivity to study hadronic interactions at energies well beyond those available in the laboratory.

Unfortunately, these issues are muddied due to the possibility of a mixed composition in the beam of cosmic rays. The most reasonable point of view with respect to this problem is to use existing accelerator data extrapolated to energies of interest to model predictions for various composition assumptions to the data. Certain EAS parameters are much less sensitive to hadronic model assumptions than others. These can be used to establish the cosmic ray composition. The other parameters can then be used to test the hadronic model predictions themselves. In this way, the measured distribution can be used to study cosmic ray composition. In this sense, the study of ultrahigh energy cosmic rays is an application of high energy physics. However, except for extreme cases, it is not presently possible to make a clean distinction between the effects of a mixed composition and a change in the properties of the hadronic interactions.

1.7.1 Ground Array Experiments

Extensive air showers were first detected by measuring the number of electrons in the EAS reaching the surface of the Earth. This original technique is still being utilized using very large (1000–3000 km²) ground arrays of detectors. These arrays are typically composed of scintillation counters or water Cherenkov tanks. The early generation of these ground arrays had areas of 1–10 km² and included: Volcano Ranch, Haverah Park, Sydney, Yakutsk, and Akeno experiments. The AGASA experiment, building on the Akeno array, reached ~100 km². Currently, the Telescope Array (TA) and the Auger array are the largest at ~3,000 km². There are also a number of smaller high-altitude arrays, used to search for point sources of PeV γ-rays, and to study cosmic ray spectrum and composition in the PeV to EeV energy range. Currently, operational arrays in this energy range include IceTop (South Pole), LHASSO (China), and Tunka (Russia).

These arrays can determine the direction from which an EAS arrives and provide an estimate of the parent particle's energy. A measurement of the transverse profile of electrons from an EAS on the surface of the Earth can be used to determine the axis of the shower. The zenith angle for an EAS can be measured by an analysis of the arrival times of the shower front impinging on the array detectors. Once the shower axis and zenith angles are known, matching the measured profile to a theoretical lateral distribution of electrons can be used to estimate the incident particle energy. In particular, the estimated particle density at a certain perpendicular distance from the shower axis is often used for this purpose. The correspondence between the lateral particle distribution or selected density and the primary cosmic ray energy is found through simulations of EAS using various high energy physics models for the primary interaction, and Monte Carlo techniques for modeling the shower.

Advantages of the ground array technique, include: good zenith angle resolution; 24/7 operation that is independent of weather; and the fact that this is a relatively cheap and stable technique that can be maintained and operated continuously over years, and even decades. The disadvantages of this technique should also be mentioned; these are the indirect nature of the energy estimate; the model dependence of the energy calibration; and the practical difficulty of getting to very large detection areas and maintaining detectors over such areas. Other indirect techniques involve detecting muons from EAS and studying the nature of the surviving hadrons near the core of the shower.

1.7.2 Atmospheric Light Emission Experiments

The newer range of techniques is based on atmospheric light emission associated with EAS. As ionizing particles in an EAS traverse the atmosphere, they produce light in two ways: Cherenkov radiation and nitrogen fluorescence. The opening angle for Cherenkov light emission in the atmosphere is approximately 2°, so the Cherenkov light is highly collimated with respect to the original particle direction. In the case of scintillation light, nitrogen molecules are excited by the EAS and can radiate their energy by producing near-ultraviolet photons. The emitted light is isotropic and, in the range between 3,000 and 4,000 Å, corresponds to 4–5 photons/m per ionizing particle, or ~20 photons/MeV of deposited energy. Near shower maximum, there is such an enormous number of ionizing particles (10^8–10^{10}) that even this low efficiency process yields measurable amounts of light. Due to the directionality implicit in the Cherenkov technique, an EAS must point towards the detector to be seen. This means that there is a reduced solid angle of acceptance. The Cherenkov technique is, therefore, most useful for lower energy events where the flux of particles is higher and, particularly, in searching for point sources, since it gives very good angular resolution. The fluorescence technique allows a large collection area because fluorescence light is isotropic and has an attenuation length of 10–20 km, which makes it useful for studying the highest energy cosmic rays.

The fluorescence technique was first demonstrated as practical by the Fly's Eye group, which then developed the High Resolution Fly's Eye (HiRes) experiment. It is the only technique that directly measures the position of the shower maximum in the atmosphere. Subsequently, hybrid techniques utilizing large surface arrays and air-fluorescence telescopes operating in coincidence were established by the Auger and Telescope Array groups. The hybrid technique reduces the energy scale uncertainty of the surface detectors by cross-calibration with air-fluorescence, as well as improving the geometrical reconstruction precision of events and also providing composition and hadronic interaction dependent measurements.

1.7.3 New Approaches to Detecting UHECR

In recent years, impressive progress has been made in using radio to detect EAS in the atmosphere (see the discussion and references in Chapter 7). The mechanism for producing radio signals in an electromagnetic cascade (the geomagnetic effect and the Askaryan effect) are well- understood, but require very careful modeling and simulation. This is now in hand. Results from prototype experiments such as LOFAR and AERA look very promising. It appears that both the shower energy and the location of the shower maximum (X_{max}) can be accurately determined. However, this approach still requires the deployment of a large number of closely spaced detectors. An alternative approach, using radio detectors on a balloon flying over the Antarctic ice, has been pioneered by the Antarctic Impulsive Transient Antenna (ANITA) collaboration.

With the onset of cosmological neutrino detectors (such IceCube) and gamma-ray detectors with sensitivity extending beyond TeV energies (such as Veritas,[13] Hess,[14] and MAGIC[15]), it now becomes possible to combine information from cosmic rays, gamma rays, and neutrinos to further define the nature of the cosmic ray accelerators. This so-called "multi-messenger" approach is in its initial phase of development, but could lead to real breakthroughs in our understanding.

References

[1] M. Tanabashi et al., Particle Data Group, *Phys. Rev. D*, **98**, 2018, p. 030001.

[2] Ibid., p. 030001.

[3] M.S. Longair, *High Energy Astrophysics*, 2nd edn, Cambridge, Cambridge University Press, 1994, p. 341.

[4] J.S. George et al., *Ap. J.* **698**, 2009, p. 1666; K. Lodders, *Ap. J.* **591**, 2003, p. 1220.

[5] G.B. Christiansen, G.V. Kulikov, and Y.A. Fomin, *Ultrahigh Energy Cosmic Rays* (in Russian), Moscow, Atomizdat, 1975, p. 5.

[6] Ibid., p. 8.

[7] E. Fermi, *Phys. Rev.*, **75**, 1949, p. 1169; H.S. Longair, op. cit., p. 344.

[8] G.B. Christiansen et al., op. cit., p. 16.

[9] G.B. Christiansen et al., op. cit., p. 9.

[10] K. Greisen, *Phys. Rev. Lett.*, **16**, 1966, p. 748; G.T. Zatsepin and V.A. Kuzmin, *JETP Lett.*, **4**, 1966, p. 78.

[11] C.T. Hill and D.N. Schramm, *Phys. Rev. D*, **31**, 1984, p. 564.

[12] F.W. Stecker, *Phys. Rev. Lett.*, **21**, 1963, p. 1016; V.S. Berezinsky and G.T. Zatsepin, *Soviet J. Nucl. Phys.*, **11**, 1970, p. 111; C.T. Hill and D.N. Schramm, *Phys. Lett.* **131B**, 1983, p. 247.

[13] J. Perkins et al., *New Astron. Rev.*, **48**, 2004, p. 637.

[14] J.A. Hinton, *New Astron. Rev.*, **48**, 2004, p. 331.

[15] J. Cortina et al., *Astrophys. Space Sci.*, **297**, 2005, p. 245.

2

EXTENSIVE AIR SHOWERS

2.1 Introduction

Direct detection of primary cosmic rays is possible for energies below 0.1 PeV (for balloon-borne experiments), ranging to ~1 PeV (for satellite experiments). At higher energies, the cosmic ray flux is too low for the limited size of an experiment that can be carried on a balloon or rocket. Thus, at higher energies one must make use of the extensive air shower (EAS) created by the interactions of the primary cosmic ray with the nitrogen and oxygen of the atmosphere.

There are several component parts to an EAS. Near the top of the atmosphere, a cosmic ray striking the nucleus of an atmospheric atom will, due to the strong interaction, produce a large number of secondary hadrons, most of which are π mesons. About one third of these particles will be π^0, which immediately decay into two γ-rays. The $\pi^\pm$ continue in the development of the hadronic component of the EAS due to the strong interaction, but the γ-rays represent the beginning of an electromagnetic component that progresses through pair production and bremsstrahlung as the EAS develops in the atmosphere. At every stage of the shower development, the production of π^0 particles continues to transfer energy from the hadronic component to the electromagnetic component of the shower. Some of the $\pi^\pm$ decay to $\mu^\pm$, creating a muonic component to the EAS. The neutrinos produced in the $\pi^\pm$ decays escape without further interactions, and represent a "missing energy" component to the EAS. Figure 2.1 illustrates the electromagnetic and hadronic processes.

EAS do not expand forever. As each component develops, more particles are created and the average energy of the particles decreases. Deep into the shower, the transfer of energy from the hadronic component to the electromagnetic component results in $e^\pm$ dominating the

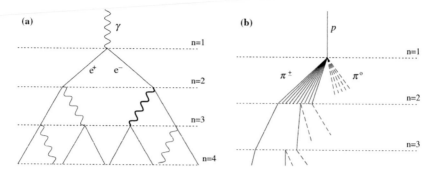

Figure 2.1 Schematic views of: (a) an electromagnetic cascade; and (b) a hadronic shower. In the hadronic shower, dashed lines indicate neutral pions which do not re-interact, but quickly decay, yielding electromagnetic sub-showers (not shown). From J.H. Kim, PhD thesis, University of Utah, with permission.

number of charged particles in the shower. The electromagnetic component then reaches a maximum when the average $e^\pm$ energy reaches "critical energy," where more energy is lost from ionization than from radiation. After this point, electrons are absorbed from the shower and the shower wanes. The depth of maximum development of the shower (called X_{max}) has an important role in the analysis of ultrahigh energy cosmic ray (UHECR) data. Since all showers reach X_{max} when average electron energies are at the same energy (the critical energy), the development of all showers near and beyond X_{max} are very similar. This constitutes a great simplification for cosmic ray physicists. As we shall see, at X_{max} showers of different energies only differ in the number of electrons, and that number is proportional to the energy of the primary cosmic ray. This fact is important for experiments using ground arrays and also fluorescence detectors.

The lateral distribution of particles in the shower is important for experiments, too. Showers have a "core" about 100 m across where almost all the particles are found. Here, about 90% of the charged particles are electrons. Moving outward from the core, the distribution of particles decreases strongly, but extends for kilometers. At distances of less than ~1 km from the core (measured perpendicular to the shower direction), electrons dominate the types of charged particle but, at greater distances, there are more muons than electrons. At all locations, photons are an order of magnitude more frequent than charged particles.

So, cosmic ray experiments covering the PeV energy range or higher make use of the characteristics of one or more of the components of EAS to reconstruct the important variables to be measured for each event: the energy, composition, and the direction of the source of the primary particle. In the remainder of this chapter, we will describe simplified models of EAS components, following J. Matthews.[1] In Chapter 3, we will describe Monte Carlo techniques for studying EAS.

2.2 Heitler Model of Electromagnetic Showers

The Heitler model for an electromagnetic shower (one initiated by a photon, electron, or positron)[2] consists of conceptualizing the shower as a series of discrete steps that occur as the shower front traverses the distance where the particles lose half of their energy. This distance is $d = x_{em} \ln 2$, where d is the distance measured in g/cm², and x_{em} is the radiation length in air (85 g/cm²). At each step, an electron or positron radiates a photon, or a photon produces an electron–positron pair. At the nth step, the number of particles is 2^n and, if the original energy is E_0, the energy of each of the particles at the nth step is $E_n = E_0/2^n$. If the critical energy is E_{emc}, then at X_{max} we have $E_{emc} = E_0/2^{nc}$, where nc is the number of steps needed to reach the X_{max} point, and the number of particles there is E_0/E_{emc}; that is, proportional to the primary cosmic ray energy.

The depth at shower maximum is another important quantity to calculate and is achieved by applying Equation (2.1)

$$X_{max} = d \ nc = d \ \ln\left(E_0/E_{emc}\right)/\ln 2 = x_{em} \ \ln\left(E_0/E_{emc}\right) \qquad (2.1)$$

from which we see that X_{max} increases logarithmically with energy. The elongation rate, λ, is defined as the rate of increase of X_{max} when E_0 changes by 1 decade; that is, $\lambda = d(X_{max})/d(\log_{10}(E_0))$. This is $\lambda = 2.3 \ x_{em} = 85$ g/cm² in air. We have to remember that this is for purely electromagnetic showers. In a cosmic ray shower, the fraction of electromagnetic particles does not start out at 100%, but grows as π^0 mesons are produced. This reduces the elongation rate.

When compared with detailed simulations of electromagnetic showers induced by photons, for example, one finds that, in the Heitler model, there are too many electrons and insufficient photons. This is due to two effects: first, several photons can be radiated per discrete step; and, second, electrons ranging out due to ionization losses are not considered in the Heitler model. A simple fix is to introduce a correction factor g (≈ 10), which we use to reduce the number of electrons in the shower at X_{max}.

2.3 Matthews–Heitler Model of Hadronic Showers

In following Matthews,[1] we again conceptualize the shower as occurring in steps of $d = x_{int} \ln 2$, where x_{int} is the interaction length of π mesons, and d is scaled to be the distance in which they lose half of their energy. We follow the development of the shower in these steps. At each step, N_{ch} charged pions are created, as are $N_{ch}/2$ neutral pions. The π^0 mesons decay immediately, transferring energy to the electromagnetic component of the shower. The charged pions continue to interact strongly, fueling the development of the hadronic component.

We recognize that there is also a critical energy for pions, $E_{\pi c}$, which is the energy where, in the next step, 50% of the pions interact and 50% decay. In the Matthews–Heitler model, once the pions reach $E_{\pi c}$ they all decay immediately. This typically occurs before the shower reaches X_{max}. So, in this model a shower has a hadronic component that is developing at a step size of about 83 g/cm² (characteristic of pion interactions in the atmosphere) while also feeding the electromagnetic component, which is developing at a step size of 53 g/cm² (which is $x_{em} \ln 2$). This mismatch in development scales means that calculations in closed form are difficult, and a computer must be used. However, we can make an estimate of X_{max}, which will prove useful.

After n atmospheric layers, there are $N_{\pi} = (N_{ch})^n$ charged pions in the shower; these have a total energy $(2/3)^n E_0$. The rest of the energy has been transferred into the electromagnetic component. The energy of individual pions in this layer is $E_{\pi} = E_0/(3N_{ch}/2)^n$. At the point of the maximum extent of the shower, the total energy of the cosmic ray consists of electrons at the critical energy and muons at the pion critical energy: $E_0 = N_{max}E_{emc} + N_{\mu}E_{\pi c}$. This is $E_0 = 0.85(N_{max} + 24N_{\mu})$, in GeV. So, in this model, if one measures the number of electrons and muons at the X_{max} point, one can calculate the total energy of the primary cosmic ray.

To estimate the elongation rate of a shower initiated by a proton, we examine the point of maximum extent of the sub-showers made by photons in the first interaction of the primary with the nucleus of an atmospheric atom. The energy of one of these photons is $E_{em} = E_0/3N_{ch}$, and its maximum extent occurs at $X_{max} = x_p + x_{em} \ln(E_0/(3N_{ch}E_{emc}))$. Here, x_p is the proton interaction length in air. The value of the elongation rate becomes $\lambda_p \approx 58$ g/cm^2. This is close to results from Monte Carlo calculations, and is very similar for proton-induced and nucleus-induced showers.

Finally, we discuss the relation between X_{max} from showers made by nuclei and those by protons. The simplest model is that of superposition; that is, in the first interaction, the nucleus breaks up into A showers of energy E_0/A, where A is the number of protons plus neutrons in the nucleus. There are two changes we must make to the X_{max} formula given in the preceding paragraph. First, x_p is reduced to x_A, the interaction length of the nucleus on air. Second, we must move down the elongation curve by a distance of $\lambda \log_{10} A$. For ^{56}Fe this, is about 100 g/cm^2.

References

[1] J. Matthews, *Astropart. Phys.*, **22**, 2005, p. 387.
[2] W. Heitler, *The Quantum Theory of Radiation*, 3rd edn, London, Oxford University Press, 1954, p. 386.

3

MONTE CARLO SIMULATION TECHNIQUES

3.1 Introduction

In Chapter 2, we discussed the general characteristics of cosmic ray extensive air showers (EAS), and learned from simple models that near X_{max} all showers follow a similar development but differ in the total number of particles (which is proportional to E), that the value of X_{max} rises logarithmically with energy, and that the difference in X_{max} between proton- and iron-induced showers is on the order of 100 g/cm^2. Here, we discuss techniques for understanding EAS that hopefully are more accurate.

There are two reasons for doing so. First, we want to gain an understanding of EAS for their own sake. Do more complete simulations bear out the results of simple models? Can we build models of EAS using quantities characterizing particle interactions measured in accelerator experiments, even if accelerators do not have sufficiently high energy? Second, we want to use the results of simulations to understand experimental results on EAS. Here, the aim is to measure the spectrum and composition of ultrahigh energy cosmic rays (UHECR), and to search for anisotropy in the directions from which they arrive. We will see that Monte Carlo simulations of EAS are very important for these studies.

In measuring the spectrum of UHECR, some detectors have 100% detection efficiency above a certain threshold energy. In particular, a surface detector (SD), which consists of scintillation counters or water tanks spread out in an array, can cover a large area and be sensitive to UHECR. But if energies are below a certain level, not all EAS will be detected. In order to measure the spectrum below the threshold energy, one must correct for the missing events; that is, calculate the efficiency of the detector at those energies. This often can be done using Monte Carlo shower simulations.

Other types of detector (e.g., fluorescence detectors: FDs) require an efficiency calculation to measure the UHECR spectrum. These detectors look out from a central station and can see the produced fluorescence light from EAS which is brighter than a certain threshold amount. Since the brightness is proportional to the cosmic ray's energy, the maximum distance from which a cosmic ray of a given energy can be seen varies with energy. Since one does not know the maximum distance from first principles, a Monte Carlo shower simulation must be used to measure the detector efficiency.

In determining the composition of UHECR, the best method conceived to date uses X_{max} as seen by an FD. Although other techniques exist, to be validated they are usually calibrated against the X_{max} technique. As we shall see in Chapter 13, there is considerable uncertainty in assigning the composition to either hydrogen or other nuclei, due, in part, to X_{max} measurement uncertainties and to uncertainties in Monte Carlo air shower programs. The latter problem centers on the extrapolation of production cross sections of pions and other hadrons from the Fermilab and CERN measurements to the (much higher) energies needed for cosmic ray experiments.

3.2 Programs to Simulate Air Showers

Considerable effort has been put into developing Monte Carlo programs to simulate cosmic ray air showers. The most commonly used shower program is called CORSIKA[1], which follows the production, interaction, and decay of particles in a cosmic ray shower; another program in use is Aires.[2] CORSIKA keeps a list of all the particles at a given point in a shower, called the "stack," remembering the particle's identity, energy, and position and momentum vectors. CORSIKA propagates each particle through the atmosphere, in steps, calculating at each step whether the particle will interact and, if so, how. The interaction probabilities and their details are provided by functions called "hadronic generators." New entries in the stack are created for particle production, and entries are removed as required; for example, when particles fall below a user-settable threshold energy level, gamma rays that undergo pair production, and so on. Since there are about 10^9 charged particles at shower maximum initiated by a 1 EeV cosmic ray, CORSIKA simulations require considerable processing time.

To reduce these processing requirements, CORSIKA uses an approximation technique called "thinning."[3] Here, after the particles in the shower have fallen below a user-settable threshold (often chosen to be 10^{-6} of the primary energy), they are considered on a statistical basis. If a particular particle is chosen to be eliminated, another at a similar point in phase space is given a weight to take that into account. The weight is another element of the stack. Thinning substantially speeds up EAS simulations, and has little impact on average values of energy and X_{max} for showers detected by air fluorescence. However, it underestimates fluctuations around average values and cannot be directly used to simulate SD response to EAS.

3.3 Hadronic Generator Routines

When CORSIKA is considering the next step in the propagation of a particle in the atmosphere, it needs to know the probability of interaction or decay. It calls up a hadronic generator routine to obtain this information. Several hadronic generators exist, the most commonly used being QGSJet,[4] Sibyll,[5] and EPOS,[6] each of which comes in different versions.

The problem, of course, is that cosmic rays of energy up to 3×10^{20} eV have been seen, but accelerator beam energies are below 7×10^{12} eV. The highest center of mass energy (13 TeV at the CERN Large Hadron Collider or LHC) is equivalent to a cosmic ray proton of energy 9×10^{16} eV striking one nucleon in an air nucleus. This may be near the bottom end of the UHECR range, but the highest energy cosmic ray observed, at 3×10^{20} eV, is equivalent to 750 TeV in the center of mass. An extrapolation in energy of a factor of 58 is required to describe how it interacted. The equations of the strong interaction have never been solved, so a rigorously correct extrapolation is impossible. The solution that has been chosen by the authors of QGSJet and EPOS is to adopt Reggeon field theory models of the strong interaction that capture the flavor of the strong interaction. The authors of Sibyll have coded a "mini-jet" model into their hadronic generator and, as one increases the energy of a cosmic ray, more mini-jets are added. The models are tuned to accelerator data, but their accuracy in extrapolation is difficult to quantify.

It is possible to use any of the models in calculating the efficiency or the resolution of a cosmic ray detector. The interpretation of X_{max} measurements, however, is much more dependent on the model used. The quantities that have the largest effect on predictions of X_{max} for hadronic generator routines are the p-p total cross section, and the multiplicity and elasticity of high energy p-p interactions. The p-p total cross section determines the starting point of showers high in the atmosphere. The multiplicity (see N_{ch} from Chapter 2), its variation from event to event, and the energy distribution of produced pions are clearly important to how all components of a shower propagate through the atmosphere, and the elasticity specifies how much energy is carried away by the leading hadron in the interaction. Two hadronic generators, QGSJet II-4 and EPOS-LHC, have tuned their cross sections to accelerator measurements going up to the LHC energy of 13 TeV in the center of mass, while many quantities of interest were measured at Fermilab fixed target energies, which reached 1 TeV in the lab frame. The energy 13 TeV, as we have seen, is equivalent to cosmic ray energies just below 1×10^{17} eV but, in the UHECR regime, say at $10^{19.5}$ eV, cross sections at 250 TeV in the center of mass are needed.

3.4 Uncertainties from Extrapolation of Cross Sections

Figure 3.1 shows the prediction of CORSIKA and five hadronic generators for the depth of shower maximum as a function of energy. The variation for protons among the models is about 25 g/cm² at 10^{17} eV, and 35 g/cm² at $10^{19.5}$ eV.

Using a technique pioneered by Ulrich et al.,[7] Abbasi and Thomson[8] estimated the uncertainty in <X_{max}> due to extrapolation at 10^{17} eV and $10^{19.5}$ eV. Figure 3.2 shows the sensitivity in X_{max} to variations in input parameters: p-p total cross section, and the multiplicity and elasticity of high energy p-p interactions. The horizontal scale, $f(E)$ is given by Equation (3.1):

$$f(E) = 1 + (f_{19} - 1)\log(E/10^{15} \text{ eV})(\log(10^{19} \text{ eV}/\log(10^{15} \text{ eV})) \quad (3.1)$$

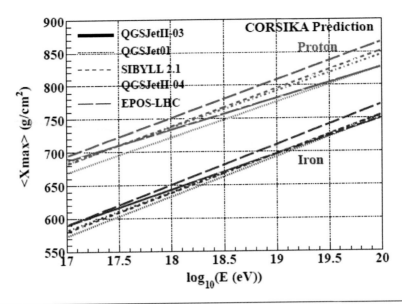

Figure 3.1 $<X_{max}>$ predictions from five hadronic models as a function of energy. From R.U. Abbasi and G.B. Thomson, arXiv:hep-ex/1605.05241, Cornell University, with permission.

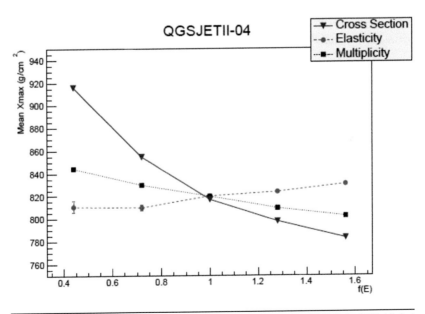

Figure 3.2 Dependence of $<X_{max}>$ on cross section, elasticity, and multiplicity for QGSJet II-4 at $10^{19.5}$ eV. The lines are drawn to guide the eye. From R.U. Abbasi and G.B. Thomson, arXiv:hep-ex/1605.05241, Cornell University, with permission.

where E is the energy, and f_{19} is the scaling factor to cosmic ray showers at 10^{19} eV. The result in Abbasi and Thomson[8] is that the uncertainty in $<X_{max}>$ due to extrapolation is ± 3 g/cm^2 at 10^{17} eV, and ± 18 g/cm^2 at $10^{19.5}$ eV. The latter spans the full range of model predictions.

3.5 Efficiency Calculation in Fluorescence Detector Analysis

The flux of cosmic rays, $J(E)$, is given by the formula $J(E) = N(E)/(\varepsilon A\Omega T\Delta E)$ where $N(E)$ is the number of cosmic rays seen at energy E, T is the observation time, ε is the efficiency of the detector at that energy, A is the area covered, ΔE is the width of the energy bin, and Ω is the solid angle subtended by the detector. The product $\varepsilon A\Omega$ is called the "aperture." Typically, the flux is measured in bins of size 0.1 in $\log_{10}(E)$ (sometimes 0.05 bins are used for regions of high statistics). As we have seen, e must be calculated by Monte Carlo simulation.

In our description of how to calculate e, we will follow the methods used in the High Resolution Fly's Eye (HiRes)[9] and Telescope Array (TA)[10] experiments. The techniques in other experiments are generically similar. The simulation starts with a large set of events generated using CORSIKA. Developing such a simulation often entails running many changes and versions of the program we are describing. Therefore, one does not want to generate a large number of CORSIKA events many times (doing this just once is very time consuming), so a library of CORSIKA showers is generated and saved, and is reused many times. In order to take experimental resolution into account, a continuous energy distribution is generated, using a spectrum as measured by previous experiments. Also, for resolution purposes, the minimum and maximum energies generated should extend beyond the relevant experimental limits. Since events at different zenith angles look different to an FD, CORSIKA events are also generated at a variety of zenith angles. An FD is sensitive to the core of the shower, not the extended tail, so the information saved for each event could be the number of charged particles (or, equivalently, the energy deposited in the atmosphere) per bin of atmospheric depth, measured in g/cm^2.

The simulation of an experiment starts by selecting an event randomly from the shower library. The position of the event in the vicinity of the detector, together with its zenith and azimuthal angles, are chosen randomly. The yield of fluorescence photons per ionizing particle or per deposited energy in the atmosphere has been measured (to about 10% accuracy) by a number of laboratory experiments, so the number of photons that would strike the detector can be calculated. The atmosphere scatters fluorescence photons, which is also taken into account. The uncertainties in this process go into the systematic uncertainty of the spectrum.

Next, the operation of the detector has to be simulated. All detectors have a "trigger"; that is, conditions that have to be satisfied before an event's information is collected and saved. For an FD, track length and brightness are often parts of the trigger. The same CORSIKA event, placed close to the FD, could satisfy the trigger but, when placed too far away, will fall short of satisfying the requirements. For example, for an event of energy 10^{18} eV, and detectors such as those of the HiRes, TA, and Auger experiments, the trigger will be satisfied if the event is within ~10 km of the FD. Higher energy events will trigger the detector from greater distances. Other items that have to be taken into account are the solid angle of the FD mirror, the reflectivity of its surface, the absorption of a UV filter in front of the photomultiplier tubes, tube quantum efficiency and gain. Finally, the output pulse of the tube is calculated, and the response of the front-end electronics to this pulse and that of other tubes of the telescope is tested against the trigger conditions. If the trigger is satisfied, the event is said to be "accepted," and information about the event is saved and written out in exactly the same format as the data.

The analysis of the Monte Carlo events is then performed using exactly the same program that is used for the data. Quantities such as the event's energy, X_{max}, and directions of the source are reconstructed. This leads to there being two values of these quantities: the thrown value and the reconstructed value. The difference between them is called the "resolution" of the experiment. At this point, all of the inputs to the set of Monte Carlo events are known quantities. If the of determination of these quantities has been done well, then histograms of important variables in the data and the Monte Carlo results should be identical. This is a crucial test, and if there are significant differences in these data–Monte Carlo comparisons, one must go back and see where any error has been made. It is the experience of all experiments that have FDs that the data–Monte Carlo comparisons are good; that is, this process converges.

After convergence, we can calculate the efficiency of the detector in each energy bin, which is $e = N_{acc}/N_{thr}$, where N_{acc} is the number of accepted events in this energy bin (measured using the reconstructed energy), and N_{thr} is the number of events in this energy bin (measured using the thrown energy) generated in the Monte Carlo process.

3.6 Efficiency Calculation in Surface Detector Analysis

The general outline of the efficiency calculation in surface detector (SD) analysis is the same as described for an FD, but there are some important differences. For an FD only the core of the shower is important; the particles in the tails of the shower do not generate sufficient fluorescence light to be visible. There are so many particles in the core that the thinning approximation used in CORSIKA works well for an FD. If there are 10^{10} particles at shower maximum, but the result of thinning is that only 10^4, for example, are used in CORSIKA, the statistical uncertainty in the shower size is $\pm1\%$, which is sufficiently good for experimental purposes.

However, the tails are the most important element for an SD because, if the core comes close to one of the SD counters and many particles strike it so that it saturates, much information is lost, and that counter is almost unusable. So, the number of particles that strike a usable SD counter is almost always low, and the CORSIKA thinning approximation causes important information to be lost. In this approximation, the mean number of particles hitting a counter is accurate, but the variation from event to event is underestimated. It is therefore impossible to determine the experiment's resolution if thinning is used. Experiments below about $10^{16.5}$ eV can use CORSIKA with no thinning, because modern computer farms have sufficient power to fulfill the processing needs. Experiments that reach higher energies have typically only used their SDs in the energy range where they are 100% efficient because of this difficulty.

There is an additional difficulty for experiments in which SD counters are water or ice tanks. All the hadronic generators that exist have one problem in common: they predict that there are too few muons in UHECR showers (of QGSJet, Sibyll, and EPOS, Sibyll has the smallest muon problem). Since a muon striking a water tank SD counter gives a higher pulse height than an electron or photon, water tanks emphasize the muonic component of EAS. In the Auger experiment, for example, ~85% of the water tank signal comes from muons. This creates serious data–Monte Carlo comparison mismatches. The spectrum measurements of these experiments use only the 100% efficiency energy regime of their detectors.

The one exception is the TA experiment, the SD of which is made of scintillation counters where electrons and muons give the same counter pulse height. Here, the muon problem is much smaller (muons contribute ~15% to counters' signals), so it may be possible to operate off the 100% efficiency plateau. But the thinning problem would remain. For this reason, the TA collaboration performs a "dethinning" step[11] in their Monte Carlo generation. The weight, w, of each particle followed in CORSIKA is turned into w particles, which are projected to the ground within a cone and are used in the simulation. Dethinning was tuned using 101 UHECR CORSIKA events generated without thinning on a large computer farm. When applied to the TA SD Monte Carlo, very good data–Monte Carlo comparison plots result. So, even though the 100% efficiency plateau starts at about 9×10^{18} eV, TA spectra measured by their SD covers the energy range above 1.6×10^{18} eV.

References

[1] D. Heck et al., CORSIKA: a Monte Carlo code to simulate extensive air showers, FZKA-6019, 1998.

[2] S.J. Sciutto, arXiv:astro-ph/9911331, Cornell University.

[3] A.M. Hillas, *Proceedings of the Paris Workshop on Cascade Simulations*, J. Linsley and A.M. Hillas (eds.), 1981, p. 39.

[4] S.S. Ostapchenko, *Phys. Rev. D*, **83**, 2011, p. 014018.

[5] E.-J. Ahn et al., *Phys. Rev. D*, **80**, 2009, p. 094003.

[6] T. Pierog et al., *Phys. Rev. C*, **92**, 2015, p. 034906.

[7] R. Ulrich, R. Engel, and M. Unger, *Phys. Rev. D*, **83**, 2011, p. 054026.

[8] R.U. Abbasi and G.B. Thomson, arXiv:hep-ex/1605.05241, Cornell University.

[9] T. Abu-Zayyad et al., *Proceedings of the 26th International Cosmic Ray Conference*, Vol. 4, 1999, p. 349; J.H. Boyer et al., *Nucl. Instrum. Methods Phys. Res. A*, **482**, 2002, p. 457.

[10] J.N. Matthews et al., *Proceedings of the 32nd International Cosmic Ray Conference*, 2011, p. 273.

[11] B.T. Stokes et al., *Astropart. Phys.*, **35**, 2012, p. 759.

EXPERIMENTAL TECHNIQUES
Surface Detectors

4.1 Introduction

One of the most important experimental techniques for studying UHECR is arrays of particle detectors spread out on the ground, often called ground arrays or surface detectors (SDs). The detectors are placed in an array, usually with array unit cells being rectangular, triangular, or hexagonal, where the detectors cover a small fraction of the area where the shower particles fall. The measurement consists of sampling the number of particles in the shower as a function of detectors' locations on the ground.

SDs are often made up of scintillation counters (which are equally sensitive to electrons, muons, and charged hadrons, but less so to photons), water tanks or ice tanks (where muons give much larger pulse heights than other particles), or scintillation counters buried underground or with thick absorber placed above them so that they are sensitive only to muons.

Each SD counter communicates with the outside world through either cables or radio. Power can come to a counter over a cable and signals taken from it to a trigger system, or the counters can have solar cells and batteries for power and radio communication to the SD trigger system.

Our discussion will center on two types of SDs, which differ in the counter spacing and hence the energy ranges they cover. The dividing line between them is about 1 EeV. At low energies the KASCADE[1] and Akeno[2] experiments have a counter spacing of ~20 m, and use the sampling to calculate the total number of electrons in the shower. This is related to the primary cosmic ray's energy through Monte Carlo calculation. Experiments to study higher energy cosmic rays must cover much larger areas because the cosmic ray flux falls steeply with energy, so the luxury of small counter spacing is financially impossible. Here the best method of determining energy is to measure the density of charged particles at a given distance, d, from the core of the shower, and relate that to energy through either Monte Carlo calculation or, for hybrid experiments, comparison with the more direct measurement of a fluorescence detector (FD). Experiments such as IceTop ($d = 125$ m),[3] TA ($d = 800$ m),[4] and Auger ($d = 1000$ m) use this method.[5]

Experiments using SDs are important for historical reasons. Before the invention of the fluorescence, Cherenkov, and radio techniques, it was the only method known. Many pioneering measurements were made this way. At present, it is still the only method useful for UHECR measurements that has 100% duty cycle. The fluorescence technique can reach the highest energies, but is only useful on clear nights when the moon is down (typically ~10% duty cycle). Cherenkov light measurements also have a low duty cycle. Both Cherenkov and radio measurements require small detector spacing, and have not been extended to UHECR energies for financial reasons.

In Chapter 2, we saw that all EAS have an important property in common: at shower maximum the electrons are at the critical energy of 85 MeV, and near and beyond this point the showers develop identically independent of energy. In the fluorescence technique one observes the shower profile calorimetrically and measures the number of electrons at shower maximum, and hence the cosmic ray energy, in a direct way. An SD measures the lateral spread of a shower, at one atmospheric depth, and if the shower has fluctuated to have its maximum extent be deeper in the atmosphere, the SD will see more particles and a higher reconstruction of the energy will result.

The energy resolution of a single FD operating by itself (which is called monocular mode), however, is not better than that of an SD. This is because an FD, operating in monocular mode, has an angular resolution in the reconstruction of the position of the shower large enough that it affects the energy resolution. If two fluorescence stations see the same event (called stereo mode), or the FD and an SD see the same event (called hybrid mode), then the angular resolution improves enough that it does not affect the energy resolution.

As we will see in Chapter 9, on anisotropy, cosmic ray particles are bent by extragalactic and galactic magnetic fields, and so do not point back to their sources. Indeed, a point source at an extragalactic distance would result in a fuzzy blob of events 5–10° in diameter even for the highest energy (~10^{20} eV) cosmic ray protons. So, for anisotropy purposes, the angular resolution of an SD would be sufficient if it were smaller than 5°. This is, in fact, the case, since the resolution of a typical SD is about 1–2°.

Even with an SD's limitations, its better duty cycle, and its ability to reach the highest energies, make an SD the detector of choice for spectrum measurements, and particularly for anisotropy searches.

4.2 Use of Scintillation Counters and Water Tanks

The Haverah Park[6] and Auger experiments use water tanks for their SD, while IceTop uses ice tanks. Highly purified H_2O is used to fill a cylindrical tank with photomultiplier tubes (PMTs) looking into the medium from above. The H_2O does not scintillate but, rather, Cherenkov light from charged particles forms a cone about their trajectories which, after reflections, reaches the PMTs. The energy of electrons and photons at ground level is typically a few tens of MeV; hence, their range in water is short and their output in Cherenkov light is small. Muons at ground level have higher energies, traverse the whole order-of-meter thickness of the H_2O layer and produce larger pulse heights. Thus, the muon signal dominates the tank signal. One of the benefits of H_2O tanks is that the medium is inexpensive, meaning that larger water tanks can be built compared to the scope of scintillation counters. This helps in sampling particles particularly near threshold. Water and ice tanks also have a higher vertical profile, which enables better detection of very large zenith angle showers. This is of particular importance for searches for neutrino induced showers where maximum sensitivity is for near-horizontal events.

On the other hand, the KASCADE, Akeno, AGASA,[7] and TA experiments use scintillation counters for their SD. Liquid scintillator is used occasionally, but plastic scintillator is more common. When a charged particle ionizes atoms in its path through the scintillation material, the ions and electrons recombine to form neutral atoms. Ultraviolet light is emitted in the recombination process. The plastic is doped with a fluor that absorbs ultraviolet light and re-emits in the blue (the preferred color for PMT sensitivity). The light is collected through an air light guide or optical fibers, and brought to the PMT.

4.3 Energy Threshold of Arrays

The signal from the PMT is amplified and its area is measured by the front-end electronics of the counter. The signal also goes to a discriminator circuit with a discrimination level tuned so that single muons are counted with 100% efficiency. Single muons, which are far from the core of low energy ($\sim 10^{12}$ eV) showers, occur at about 200 Hz/m². A special trigger collects these single-muon events for calibration purposes. The counter signal is also sent to the main trigger for collecting cosmic ray events.

In most cases, this trigger requires several adjacent counters to be hit within a certain time window. For example, the TA SD trigger requires three adjacent counters to be hit within 8 µs. Consider a high energy shower the core of which strikes the ground in the middle of a square (or perhaps a triangle or hexagon) of SD counters. If the energy is sufficiently high, the nearby counters are struck by many particles, the trigger is satisfied, and the event's information is collected. However, a low-energy cosmic ray may produce too few particles in its shower, meaning that too few counters will be struck by shower particles. This event will not trigger the SD array. When the fraction of events collected falls below 100%, the SD is said to be inefficient. Figure 4.1 shows the efficiency as a function of energy for the TA SD.

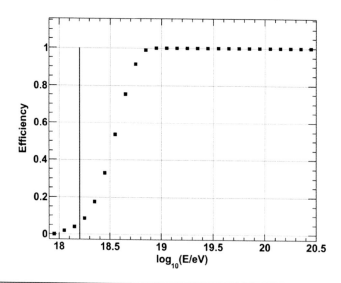

Figure 4.1   The efficiency of the TA SD as a function of cosmic ray energy. Above about $10^{18.9}$ eV, the efficiency is 100%, and is said to be "on plateau." The "knee" of the plateau curve is said to be at $10^{18.9}$ eV. The TA SD spectrum uses energies above the vertical line at $10^{18.2}$ eV. From the author, D. Ivanov, in T. AbuZayyad et al., *Ap. J. Lett.* 768, 26 (2012), with permission.

The energy where the rising efficiency reaches the 100% efficiency plateau is called the "knee" of the plateau curve. For TA, it is $10^{18.9}$ eV; for Auger, the knee is $10^{18.4}$ eV. The difference is largely due to the difference in SD counter sizes (see Section 4.4).

4.4 Introduction to Arrays Studying $E > 1$ EeV Cosmic Rays

In the Auger SD, there are about 1600 water tanks, each of 10 m² area, separated by 1.5 km in a hexagonal array covering 3000 km². This array is 100% efficient for energies above $10^{18.4}$ eV, and no attempt is made to determine the spectrum at lower energies. The Auger collaboration finds that there is a serious mismatch (~50% difference) between the number of muons seen in their detectors and the number predicted by hadronic generator programs (there is a muon shortage in the programs' predictions). This creates problems for the Monte Carlo simulation of their SD; hence, they do not calculate their SD efficiency to measure the spectrum at lower energies.

Cosmic ray air showers appear differently to an SD depending on the zenith angle of the shower. This is because the slant depth of atmosphere the shower has traversed varies approximately as the secant of the zenith angle and, when seen at ground level, a shower with a higher zenith angle has developed further. To correct for this effect, the Auger collaboration uses a method called "constant-intensity-cuts." The flux of actual cosmic rays is independent of the zenith, so one can determine the relative energies of events by associating in energy events at different zenith angles that have the same flux. This only determines the energy up to an overall constant. This constant is measured by the Auger collaboration using hybrid events (those seen and reconstructed by both FD and SD systems), since the calorimetric FD energy measurement is independent of the zenith angle.

TA, also a large hybrid experiment, has an SD of 507 scintillation counters that covers 700 km²; each counter has an area of 3 m² and they are separated by 1.2 km. The array is 100% efficient above $10^{18.9}$ eV and uses a Monte Carlo efficiency calculation to measure the spectrum down to $10^{18.2}$ eV. Since scintillation counters do not overemphasize the muonic component of air showers, the TA collaboration was able to construct an accurate Monte Carlo simulation for the TA SD. As described in Chapter 3, this construction is tested by making histograms of data and Monte Carlo events for many variables of

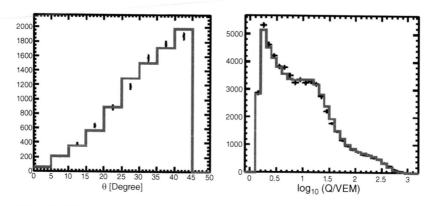

Figure 4.2   Two data–Monte Carlo comparison plots from the TA SD data. The left-hand panel shows a histogram of the zenith angle, θ, of events, and the right-hand panel shows the pulse height in SD counters in units of vertical equivalent muons (VEM). In both panels, the black points are the data and the histogram is the Monte Carlo prediction. From the author, D. Ivanov, in T. AbuZayyad et al., arXiv:1403.0644 [astro-ph], with permission.

interest. The agreement between data and Monte Carlo is used to determine the accuracy of the Monte Carlo simulation. Figure 4.2 shows two such data–Monte Carlo comparison plots from the TA SD. The agreement between the two is excellent and, from these plots— and many others—one can conclude that the Monte Carlo simulation is quite accurate.

4.5 Reconstruction of UHECR Events

The data produced by an SD sensitive in the $E > 1$ EeV range usually consist of flash analog to digital conversion (FADC) traces of struck counters. A FADC digitizes the area under the PMT pulse in short time intervals (for the TA SD, the period is 20 nanoseconds). The basic information used in the analysis is the time the pulse started and its total area. The reconstruction of an event's direction of source and energy consists of performing two fits to the data: one for the time that counters were hit, and one for the pulse heights. The left-hand panel of Figure 4.3 shows a map of struck counters for an event, where the size of the circle is proportional to the log of the counter pulse height, and the counter time is shown in color. The arrow is the projection on the ground of the event direction, and the black star is the position of the shower core. The upper right-hand panel shows the

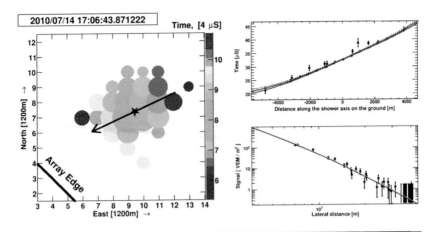

Figure 4.3   Map, time fit, and pulse height fit for an event in the TA SD data. The left-hand panel shows the positions of the struck counters in 1200 m units. The circle sizes are proportional to the log of the counter pulse area, and the color shows the time each counter was struck. The upper right-hand panel shows the time fit, in time vs. distance along the shower axis. The lower right-hand panel shows the pulse height fit, in pulse height in VEM units vs. lateral distance from the core. From the author, D. Ivanov, with permission. See www.icrc2019.org/presentations.html, Session CR17.

time fit for that event, where the time is on the vertical axis and the horizontal axis is the distance from the core position, measured parallel to the projection of the shower direction. The red line is the fit that determines the direction of the source of the shower. The lower right-hand panel shows the pulse height in the counters (in VEM, or Vertical Equivalent Muon, units, which, for TA, is 2.2 MeV energy deposit in a counter) vs. the lateral distance of the counter from the core, again measured parallel to the projection of the shower direction. The pulse height at $d = 800$ m, called "S800," is measured from the fit (the red line), and is used in the energy determination.

The energy in the TA experiment is determined in two ways. The first uses the TA Monte Carlo. A look-up table, represented graphically by the "rainbow plot" shown in Figure 4.4, is made from Monte Carlo events' energy, S800, and the zenith angle. In this figure, each colored line represents one energy, and shows how Monte Carlo events vary in S800 and sec (θ). This yields a first estimate of the event's energy. However, on comparing the first estimate with FD results for hybrid events, the TA collaboration finds that the estimate is high by 27%. This is true for all energies. Therefore, the energies determined by Monte Carlo are divided by 1.27.

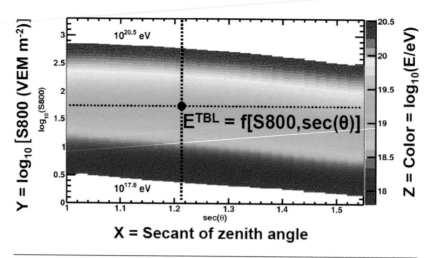

Figure 4.4 Monte Carlo determined "rainbow plot" representing the relation between the energy of events (shown as colored lines) and their pulse height and zenith angles. From the author, D. Ivanov, with permission.

The second energy determination method used in TA is that of constant-intensity-cuts. Since the Monte Carlo method might be model-dependent, the constant-intensity-cuts method is used as a check. The TA collaboration finds that the two methods agree very well. Figure 4.5 shows

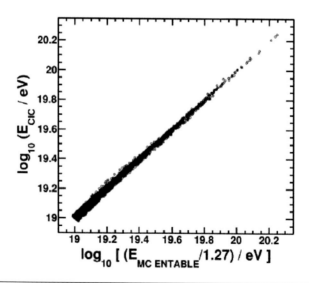

Figure 4.5 Scatter plot of energies measured by the two methods used by TA. The horizontal axis shows the Monte Carlo determined energy, and the vertical axis the energy determined by the constant-intensity-cut method. The two methods are in good agreement. From the author, D. Ivanov, with permission.

a scatter plot of energies determined by the two methods. The agreement is excellent, with a mean ratio of 0.8% and a standard deviation of 3%.

4.6 Introduction to Arrays Studying 1 PeV < E < 1 EeV

In the lower energy region, we take the KASCADE experiment as our example. KASCADE covered an energy range from about 0.5–100 PeV, and was designed to study the "knee" of the cosmic ray spectrum (which is at about 4 PeV). The energy range was later extended upward to 1 EeV by the experiment's extension, called KASCADE-Grande. The original KASCADE experiment's SD had 252 detector stations, each with both electron and muon detectors, covering 1.3% (electron) and 1.5% (muon) of the experiment's 0.04 km² area. With this coverage, sampling statistics did not affect the energy resolution; however, intrinsic shower fluctuations did. Each detector station had a liquid scintillator electron counter above 10 cm Pb and 4 cm Fe absorber, below which a plastic scintillator counter detected muons.

The KASCADE-Grande experiment consisted of KASCADE plus 37 stations of scintillation counters, of 10 m² area, separated on average by 137 m, covering 0.5 km². The efficiency of the two parts of the experiment, calculated by Monte Carlo simulation, is shown in Figure 4.6.

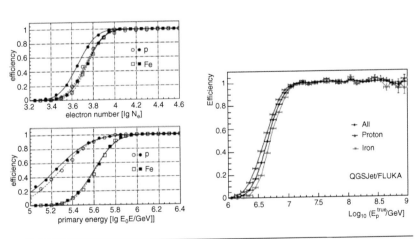

Figure 4.6 The efficiency of the KASCADE experiment (left-hand panel), as a function of reconstructed electron number, N_e, and cosmic ray energy. The filled circles are the trigger efficiency and closed circles the reconstruction efficiency, for protons and Fe. The knee of the efficiency curve is ~0.5 PeV for protons and ~0.8 PeV for Fe. The right panel shows the efficiency of the KASCADE-Grande experiment as a function of energy. The knee of the efficiency plateau is ~20 PeV. Left-hand panel from T. Antoni et al., *Nucl. Instr. Meth.*, 513, 2003, p. 490 and right-hand panel from A. Haungs et al., *Nucl. Phys. B*, 196, 2009, p. 80, with permission.

References

[1] T. Antoni et al., *Nucl. Inst. & Meth.*, **513**, 2003, p. 490. For KASKADE-Grande, see A. Haungs et al., *Nucl. Phys. B*, **196**, 2009, p. 80.

[2] T. Hara et al., Proceedings of the 16th International Cosmic Ray Conference, 1979, Vol. **8**, p. 135.

[3] R.U. Abbasi et al., *Nucl. Inst. & Meth.*, **700**, 2013, p. 188.

[4] J.N. Matthews et al., *Proceedings of the 32nd International Cosmic Ray Conference*, 2011, p. 273.

[5] I. Allerkotte et al., *Nucl. Inst. & Meth.*, **586**, 2008, p. 409.

[6] R.M. Tennant, *Can. J. Phys.*, **46**, 1968, p. S1.

[7] N. Chiba et al., *Nucl. Inst. & Meth.*, **311**, 1992, p. 338.

[8] T. Antoni et al., *Nucl. Instr. Meth.*, **513**, 2003, p. 490; A. Haungs et al., *Nucl. Phys. B*, **196**, 2009, p. 80.

5

EXPERIMENTAL TECHNIQUES
Air Fluorescence

5.1 Introduction

The idea of using air fluorescence to detect the passage of an extensive air shower (EAS) through the atmosphere occurred independently to Greisen[1] in 1960, and to Delvaille et al.[2] and Suga[3], and Chudakov[4] in 1962. An unsuccessful attempt to detect such signals was made at Cornell University in 1965 by a group led by Greisen.[5] A parallel effort by Suga and his group in Japan may have resulted in the first observation of a cosmic ray by air fluorescence.[3] The first demonstratively successful detection was achieved in 1976 by the Utah group[6] led by Bergeson and Cassiday, operating optical detectors in coincidence with the Volcano Ranch surface detector (SD) array operated by John Linsley in Arizona. This was the first "hybrid" observation of an EAS using air-fluorescence and surface array, and was the first attempt to inter-calibrate the two techniques. A complete Fly's Eye detector utilizing the air-fluorescence technique was installed at Dugway Proving Ground in Utah and began to take data in 1982.[7]

The technique relies on the fact that an ionizing particle can excite N_2 molecules in the atmosphere. Such excited molecules can de-excite by emitting fluorescence photons (typically within 10 to 50 nanoseconds after excitation). Most of this optical fluorescence comes from the 2P band system of molecular nitrogen and the IN band system of the N_2+ molecular ion.[8] A great advantage of this technique is that the fluorescence efficiency can be measured in the laboratory. In recent years, a number of experiments using accelerator beams and radioactive sources have been performed, and the fluorescence efficiency per ionizing particle is known to about 10%.[17] The measured fluorescence spectrum[9] is shown in Figure 5.1. Most of the light is emitted between 3000 and 4000 Å, which happens to be a wavelength band for which

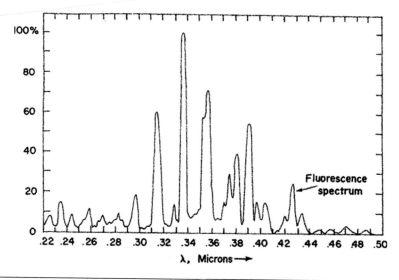

Figure 5.1   Schematic spectrum of nitrogen fluorescence in the near ultraviolet. More precise modern measurements can be found in Arqueros et al.[17] for example.

the molecular atmosphere is quite transparent, corresponding to an attenuation length of approximately 15 km for a vertical beam of light.

The fluorescent yield per particle (~5 photons/m or 20 photons/ MeV) is small but, due to the interplay between radiative and

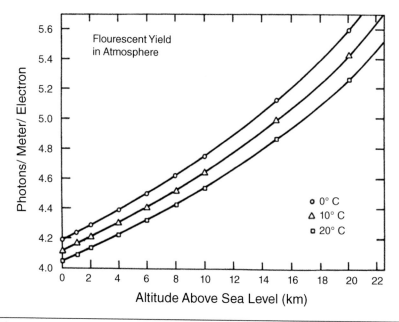

Figure 5.2   Temperature and altitude dependence of nitrogen fluorescence yield.

collisional de-excitation, only mildly dependent on altitude and atmospheric temperature[10] (see Figure 5.2). Since EAS for greater than 0.1 EeV primary energy have more than 10^8 electrons at shower maximum, a substantial number of photons is expected even with the ~.5% fluorescence efficiency.

5.2 The Fly's Eye Concept

It is instructive for pedagogic reasons to consider the original University of Utah Fly's Eye detector design in some detail, as all later air-fluorescence detectors use basically the same approach. The detector consisted of 880 phototubes in 67 mirrors 1.5 m in diameter. Each phototube is pointed in a different and unique direction, and subtends its own unique solid angle of the sky (see Figure 5.3). The entire hemisphere of the sky was imaged by these tubes. Isotropically emitted fluorescence light from an EAS was detected by those tubes whose solid angle view intersects the EAS. The relative time of arrival of this light, as well as the total integrated light, were recorded for each tube.

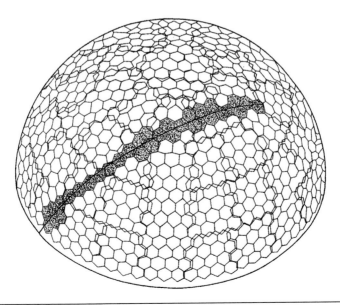

Figure 5.3   Fly's Eye phototube apertures. The shaded region represents light from the EAS striking the detector. The solid line indicates the EAS trajectory across the sky.

5.3 Signal-to-Noise Considerations

How well such a detector works depends almost entirely on signal-to-noise considerations. The problem is not so much that the signal is small (typically, with mirrors and tubes of this size, one sees $\approx$ 500–1000 photoelectrons from an EAS) but, rather, that the signal is seen in the presence of sky noise. In other words, the problem is that of attaining sufficient contrast between the EAS track and the background.

Background light contributions to night sky noise come from starlight, diffuse radiation from the galaxy, sunlight scattered by interplanetary matter, photochemical atmospheric light, and man-made light pollution. Steady state (DC) light levels *per se* do not mimic signals in the detector, but fluctuations in the DC level can. Another source of background noise is copious low energy cosmic rays, which generate Cherenkov light pulses that result in a rain of very fast pulses. The total background corresponds to $\approx 5 \times 10^5$ photons/m^2 sr µs for a wavelength interval between 3000 and 4000 Å.[11] This number can fluctuate by a factor of 2 over a given night and increases dramatically during astronomical twilight. The noise contribution can be expressed by the formula given in Equation (5.1):

$$N_{noise} = \sqrt{(4EAB\Delta\Omega\Delta t)} \qquad (5.1)$$

where $\Delta\Omega$ is the solid angle of a single photomultiplier tube, B is the overall background light, A is the collecting area of the mirror, E is the overall optical efficiency for converting photons to photoelectrons, and Δt is the integration time.

The signal, assuming a pure molecular atmosphere (see Figure 5.4), is given as shown in Equation (5.2):

$$N_{ph} = N_e N_\gamma \left((1+\cos\theta)/\sin^2\theta\right)(A/4\pi r^2)(\exp -r/\lambda_R)(c\Delta t) \qquad (5.2)$$

where N_e is the number of electrons in the EAS in the field of view of the phototube, N_γ is the photon yield per electron for atmospheric scintillation, and λ_R is the Rayleigh scattering length for light in the atmosphere (see Chapter 14); r and θ are the distance of the EAS segment viewed from the phototube and the viewing angle with respect to the shower axis. Note that $R_p = r \sin\theta$ is the impact parameter to

SHOWER GEOMETRY

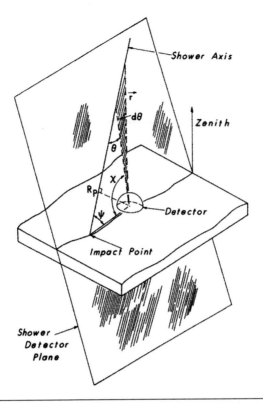

Figure 5.4   Reconstruction geometry. Once the shower detector plane is determined, the remaining variables to be determined are R_p and ψ. The shaded area $d\theta$ represents the field of view of a tube.

the axis of the EAS and, hence, we can write the signal to noise ratio as shown in Equation (5.3):

$$S/N = \left(cN_eN_\gamma\left(1+\cos\theta\right)/4\pi Rp^2\right)\left(\exp-r/\lambda_R\right)\left[EA\Delta t/B\Delta\Omega\right]^{1/2} \quad (5.3)$$

To optimize, we want to increase E, A, and Δt, and minimize $\Delta\Omega$.

For any R_p ranging from 1 to 20 km, Δt, expressed as the time it takes an EAS to cross the field of view of a tube, can range from 50 ns to 10 μs. The original Fly's Eye electronics, as well as the HiRes I detector of the next generation High Resolution Fly's Eye experiment, used sample and hold electronics to integrate the signal. In this case, once the rise of the signal voltage exceeds a preset threshold, an integrate gate is opened and the signal plus sky noise and pedestal are integrated by an analog to digital converter (ADC). The preset threshold is generally adjusted so that the singles rate for each tube is not so high as to overload the data acquisition system, typically corresponding to a fluctuation of background noise of 3 sigma from the average. The threshold can be adjusted as the night sky noise increases or decreases. The pedestal and sky noise are measured by opening the gate at preset times during the data taking period. In the case of HiRes II, and Auger and TA detectors, the signal plus background for each tube is continuously digitized by a flash ADC and stored in local memory. A subsequent circuit searches for deviations from the ambient noise and stores segments of the data stream that may correspond to actual signals. The pedestal and background are found from data just before or just after the signal.

5.4 Triggering

For the case of sample and hold electronics, the readout is initiated when a minimum number of adjacent tubes (typically five) trigger within a preset gate width. The trigger interrupts a computer, and relative times and pulse height integrals for those tubes that fired within the trigger gate are read out. Integrals are converted to photoelectrons, and the gains and reflectivity of the system are measured on a nightly basis. Similar or more sophisticated readout schemes incorporating pattern recognition can be implemented if flash analog to digital (FADC) electronics are used.

5.5 The Event Plane

Since each tube subtends a specific solid angle of the sky, an EAS trajectory appears as a track propagating along a great circle projected on the celestial sphere. This great circle determines the shower-detector plane (see Figure 5.5). The unit normal to this plane is determined by fitting a plane to all the direction vectors of tubes that detect scintillation light. If the variables R_p (the impact parameter from detector to shower axis) and ψ (see Figure 5.4) in

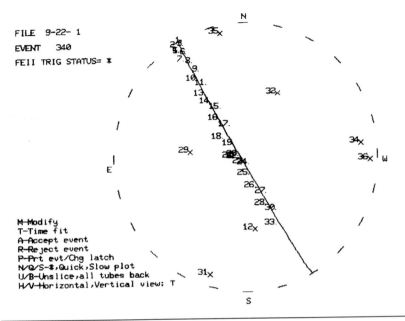

Figure 5.5   Typical event showing phototubes triggered by light from the EAS.

this plane can be determined, the complete geometry of the EAS track will be known.

5.6 The Time Fit

R_p and ψ can be determined from the relative timing of light pulses arriving in different tubes. The expected timing sequence can be calculated, since the EAS propagates in a straight line with the speed of light, and the resultant fluorescence light is emitted isotropically. The relationship between χ_i, the angle of the ith tube in the shower-detector plane, and the arrival time of light at that tube (t_i) is given by Equation (5.4):

$$\chi_i(t_i) = \pi - \psi - 2\tan^{-1}\left(c(t_i - t_0)/R_p\right) \qquad (5.4)$$

Since a large number of χ_is and t_i are typically measured for a given EAS, a best fit to the observed $\chi_i(t_i)$ of this function yields R_p and ψ, as well as their estimated errors (see Figure 5.6). This, together with the shower-detector unit normal, completely specifies the EAS trajectory, and determines the zenith and azimuthal angle of the EAS.

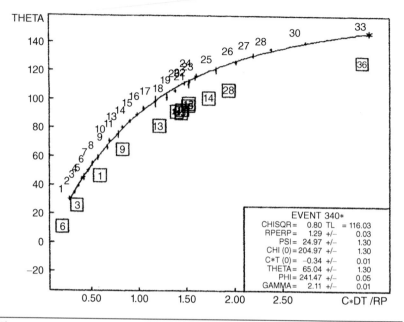

Figure 5.6 Typical event timing curve. The solid line is the result of fit to relative timing of Equation (5.4).

5.7 The Stereoscopic Method

Since there are no constraints on the timing fit, systematic errors can be a problem. The fit also requires a certain amount of curvature in the angle–time curve and this will not work well for short tracks or certain geometries. A considerably simpler and more precise technique for reconstructing the EAS trajectory was first implemented by the Fly's Eye group. A second Fly's Eye, with 36 mirrors, approximately 3.5 km from the first detected EAS seen in stereo. Such events can have their geometry reconstructed very simply, since the light from the EAS must come from the intersection of the two shower-detector planes for the two eyes (see Figure 5.7). This technique does not use timing at all (except for establishing a coincidence between the two detectors), so the results can be compared to the timing fit and systematic effects studied.

This stereoscopic method is found to be considerably more accurate than the timing method and has fewer systematic uncertainties (when the opening angle between the two shower planes is reasonable large). However, it results in a smaller collection area for showers, since both detectors must satisfy minimum trigger requirements.

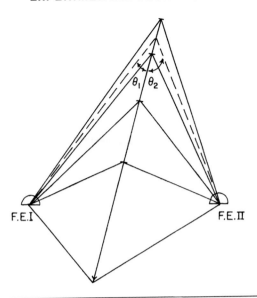

Figure 5.7 Reconstruction geometry for EAS seen in stereo. θ_1 and θ_2 are emission angles with respect to Fly's Eye I and II.

Stereo reconstruction was most importantly used in the HiRes experiment where two detector sites, HiRes I and HiRes II were separated by ~12 km. The combination of smaller phototube size (1 × 1° sky coverage vs. 5 × 5° for the Fly's Eye), larger mirrors, and larger stereo detector spacing led to a significant increase in stereo aperture at energies above 1 EeV.

5.8 Hybrid Reconstruction

Another approach to minimizing the non-linear fitting problems of monocular Fly's Eye reconstruction is to add a constraint to the fit. A surface array located in the vicinity of an air-fluorescence detector can provide such a constraint, since it measures the time of arrival of the shower front at the surface and the core location of the shower.

The distribution of pulse heights of SD counters corresponds to the shower lateral distribution and a fit to this known function can be used to establish the shower core location. The determination of shower zenith angle from surface detector timing differences can also be used to determine the zenith angle. A combined chi-square minimization can be performed to find the simultaneous best fit for the monocular fluorescence detector (FD) $\chi2$, SD core location $\chi2$, and SD zenith angle $\chi2$ data, as shown in Equation (5.5):

$$\chi2_{tot} = \chi2_{core} + \chi2_{zenith} + \chi2_{mono} \qquad (5.5)$$

More generally, a global fit can be performed using a simulated shower to match, simultaneously, the air fluorescence tube timing and amplitude, and surface detector timing, and amplitude measurements. In such an approach, the shower geometry and primary particle energy, and the position of the shower maximum all emerge from the same analysis.

5.9 Longitudinal Shower Size Determination

Once the geometry is fixed, the distance between the tubes and the EAS segments viewed by the tubes is known, so the size of the EAS as a function of atmospheric depth can be determined by inverting Equation (5.2) in Section 5.3 for N_e. The situation is actually more complex since there are a number of other sources of light:

1. Direct Cherenkov light produced by EAS particles aimed at a tube;
2. Cherenkov light scattered out of the intense light beam that is generated parallel to the shower axis via Rayleigh scattering by atmospheric molecules; and
3. Cherenkov light scattered by atmospheric impurities (aerosol scattering) (see the Appendix).

All these contributions to the total light incident on a phototube depend on the complete previous history of the shower and are not proportional to the local shower size; their contribution must therefore be subtracted out before the real longitudinal development of the EAS can be seen.

For emission angles of less than $25°$ relative to the EAS axis, directly beamed Cherenkov light dominates the light seen by the Fly's Eye detector. However, as the Cherenkov component builds up with the developing shower front, the resultant intense beam can also generate enough scattered light at low altitudes and larger emission angles to compete with the locally produced scintillation light from the exponentially decreasing shower (see Figure 5.8).

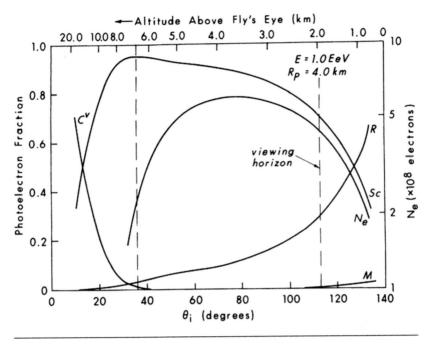

Figure 5.8 Contributions of fluorescent (N_e), direct Cherenkov (C^v), and scattered Cherenkov (R, M) to the total (S_c) light as a function of emission angle (θ_i) for a typical longitudinal shower profile.

If, for a particular geometry, ε is the scintillation efficiency, α is the direct Cherenkov efficiency per ionizing particle, and γ is the efficiency for detecting scattered Cherenkov light by either the Mie or Rayleigh process per Cherenkov beam photon, then the number of photons seen by the ith tube in the detector is given by Equation (5.6):

$$N_i = \varepsilon_i S_i + \alpha_i S_i + \gamma_i B_i \qquad (5.6)$$

where S_i is the shower size in the field of view of the ith tube, while B_i is the intensity of the Cherenkov beam propagating along the shower axis in the field of view of the ith tube. For tubes that view the beginning of the shower, the magnitude of the propagating Cherenkov beam is small, so that a good approximation would be as presented in Equation (5.7):

$$S_1 = N_1/(\varepsilon_1 + \alpha_1) \qquad (5.7)$$

Once S_1 is known, the Cherenkov beam produced by this segment of the EAS can be estimated; hence, B_i can be determined by the

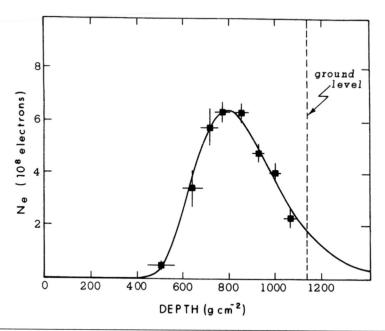

Figure 5.9   Typical reconstructed shower size as a function of atmospheric depth.

previous history of the shower. Once B_i is known, the size S_i at each tube can be determined since ε, α, and β are given by the EAS geometry plus a knowledge of the scintillation and Cherenkov light production mechanisms. Figure 5.9 shows a typical reconstructed shower size distribution. For events with large R_p, the EAS track is typically viewed at large emission angles, so that the direct Cherenkov light contribution becomes small.

A more sophisticated and accurate approach is to match the measured shower profile with simulated showers with the same geometry. The best fit simulated EAS can then predict the relative contributions of direct fluorescence to Cherenkov light.

5.10 Determining the Shower Energy

Once the size as a function of depth is known, a longitudinal shower profile function such as the Gaisser–Hillas form can be fitted to the data. The energy of the shower is then given by Equation (5.8):

$$E = (\varepsilon_0 / X_0) \int N_e(X) dX \qquad (5.8)$$

where the constant in front of the integral, with ε_0 as the critical energy and X_0 as the radiation length, is prescribed by standard electromagnetic cascade theory.[12] For a purely electromagnetic cascade, this corresponds to an average energy loss per particle of 2.18 MeV/g/cm². This number has been checked by integrating the energy loss over the energy distribution of electrons in an EAS as calculated by Hillas.[13] This calculation leads to an average energy loss per particle of 2.24 MeV/g/cm². Contemporary EAS simulation programs using proton and Fe hadronic shower models in CORSIKA lead to ~2.45 MeV/g/cm² (see Figure 5.10). Note that while this energy loss depends on the age of the shower, essentially all air-fluorescence measurements are only sensitive to shower development ages where the energy loss is essentially constant.

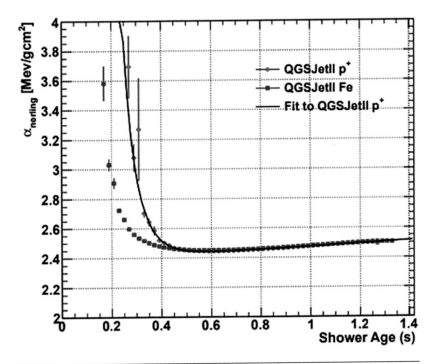

Figure 5.10 Energy loss in MeV/g/cm² as a function of shower age as simulated in CORSIKA using the QGSJet model for protons and Fe. From Elliot Barcikowski, U. of Utah Ph. D. Thesis, with permission.

5.11 Undetected Energy Corrections

The total energy found in this way corresponds to the electromagnetic energy deposited in the atmosphere. It needs to be corrected for undetected energy. Undetected energy can come from three sources:

1. Neutral particles that do not decay into charged particles before reaching the ground (neutrinos, for example);
2. High energy muons that lose most of their energy in the Earth;
3. The nuclear excitation of nitrogen by hadrons (this energy is not converted to scintillation light).

Estimates of the undetected energy as a function of the primary cosmic ray energy have been made by a number of authors. Originally, parametrization by Linsley[14] estimated corrections of approximately 13% at 10^{17} eV and about 5% at 10^{19} eV. The CORSIKA simulation package now generates missing energy corrections that depend on the hadronic model (QGS-Jet, Sybil, etc.) and range from ~10% to 5% as the primary energy varies from 10^{18} to 10^{19} eV. The dependence on p/Fe composition assumption is also about 5%.

5.12 Calculating the Differential Cosmic Ray Spectrum

Given this measured distribution of cosmic rays with energy, the next step is to find $J(E)$, the differential cosmic ray spectrum. As discussed in Chapter 3, Section 3.5, acceptance by Fly's Eye types of detector varies as a function of energy, so it must be determined using a Monte Carlo program. Here, we repeat the main points of that section for convenience.

In general, the differential spectrum is given by Equation 5.9:

$$J(E) = (dN/dE)/(t\ A\Omega(E)) \qquad (5.9)$$

where $A\Omega(E)$ in km²sr is the acceptance for events with energy between E and $E + dE$, t is the exposure time, and dN/dE is the observed distribution of cosmic rays per energy interval.

$A\Omega(E)$ is calculated in a Monte Carlo fashion in the following steps:

- An isotropic cosmic ray flux is assumed. A depth of first interaction is chosen according to an interaction length of ~70 g/cm².
- As soon as the particle interacts, a shower profile is generated using a parametrization of the simulated or real data based on

the Gaisser–Hillas form. Scintillation and Cherenkov light photons can be generated from this shower profile according to the known efficiencies for their production.

- Knowing the geometry of the Fly's Eye phototubes, one then determines which tubes will see this light and propagate the light through the atmosphere, putting in the proper atmospheric absorption coefficients.
- Given mirror reflectivities and tube quantum efficiency, one then calculates the number of photoelectrons that each of these tubes will see.
- The response of the electronics that read out the phototube current is modeled, and an estimate of whether the particular channel will trigger is made.
- The measured time of arrival of the light and the integrated pulse height corresponding to the light pulse arriving at the phototube are then calculated.
- This "fake" data is passed through the data reconstruction programs that are used to analyze data, and dN/dE, the distribution of fake events as a function of energy, is generated.

Since $J(E)$ is known in this case, the ratio of $N(E)$ to $J(E)$ gives the acceptance, $A\Omega(E)$. Figure 5.11 shows the resultant acceptance of the

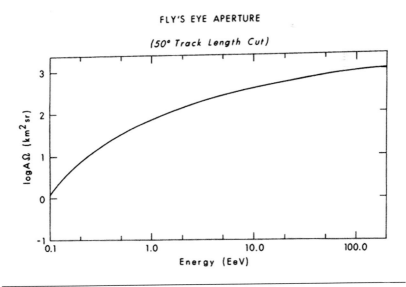

FLY'S EYE APERTURE

(50° Track Length Cut)

Figure 5.11   Fly's Eye aperture in km²-sr as a function of shower energy.

detector as a function of energy. This is a complex calculation, but it can be checked by comparing the impact parameter distribution, zenith angle distribution, and the azimuthal angle distribution between Monte Carlo and real data. Good agreement is found and this gives confidence that the "fake" data calculation is a true representation of the real acceptance of the detector.

5.13 Measuring Cherenkov Light Angular Distribution

In addition to providing more precise geometrical reconstruction, stereo data allows a check on the Cherenkov and scintillation parameters used. For example, if a segment of track is viewed by both eyes, one at a large emission angle and the other at a small angle, the ratio of observed light intensities (after correction for atmospheric attenuation and solid angle effects) is approximately the ratio of the direct Cherenkov to scintillation efficiency. Also, a track segment that is viewed by both eyes with small emission angles can be used to check the expected angular distribution of direct Cherenkov light and provide a measurement of θ_0, the Cherenkov angular distribution parameter (see Equation 5.10), since:

$$L_1/L_2 = \left[\exp\left(-(\theta_1-\theta_2)/\theta_0\right)\right]\sin\theta_2/\sin\theta_1 \tag{5.10}$$

Here, L_1 and L_2 are the corrected light intensities, and θ_1 and θ_2 are the emission angles. Figure 5.12 shows the result of such a study, and yields $\theta_0 = 4.0° \pm 1.2°$ for an average E_{min} of 34 MeV.

It is also possible to find classes of events where the signal is dominated by Cherenkov light and determine an energy spectrum the energy scale of which is based on the very well-established physics of Cherenkov light production. This can then be compared, in the same energy range, to a spectrum based on air-fluorescence as a cross-check of the air-fluorescence efficiency. Reconstructing events using Cherenkov light also extends the low energy threshold, since Cherenkov light will dominate at low energies. The Telescope Array Low Energy Extension (TALE) detector has demonstrated that spectra can be measured to below 10 PeV in energy. This technique may make it possible to overlap with direct cosmic ray measurements at, or just below, the knee of the spectrum in the near future. Such an

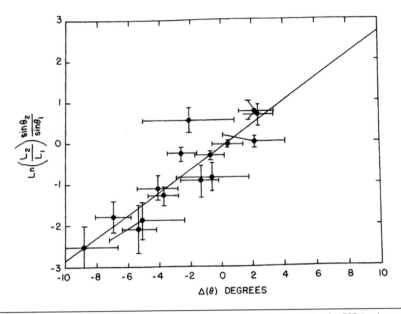

Figure 5.12 Ratio of corrected light yield to difference in emission angles for EAS track segments viewed in stereo. The solid line represents best fit to data.

overlap would be of great importance in constraining and calibrating the indirect air-shower technique of determining composition with direct elemental composition measurement.

5.14 Improvements to the Basic Technique

At the beginning of this chapter, we demonstrated that the sensitivity of the Fly's Eye depends on the ratio of tube aperture to mirror collection area. Reducing tube apertures to 1° and increasing mirror diameters to ~2 m (as is the case for HiRes[18], Auger[19] and TA[20] fluorescence detectors), yields an increase in signal-to-noise ratio of a factor of five. The new generation of air fluorescence detectors can trigger on EAS as far away as 40 km while still allowing very detailed sampling of the longitudinal development. The increased collecting area implies an order of magnitude increase in the number of EAS detected above 10 EeV per year. The use of FADC digitization of signals also improves the signal to noise ratio S/N, as it effectively allows setting an integrate gate-width that more nearly matches the signal width. Modern electronics allows much more sophisticated triggering schemes and more precise monitoring of sky-noise backgrounds variation, resulting in greater sensitivity and precision.

References

[1] K. Greisen, *Ann. Rev. Nucl. Sci.*, **10**, 1960, 63.

[2] J. Delvaille et al., *J. Phys. Soc. Jpn.*, **17**(Suppl. A-III), 1962, 76.

[3] K. Suga, *Proceedings of the 5th Interamerican Seminar on Cosmic Rays*, La Paz, Bolivia, 1962, vol. **II**, XLIX; G. Tanahashi et al., *Proceedings of the International Conference on Cosmic Rays*, Budapest, 1969, EAS 4/3, 24.

[4] A.E. Chudakov, *Proceedings of the 5th Interamerican Seminar on Cosmic Rays*, La Paz, Bolivia, 1962, **vol. II**, XLIX.

[5] L.G. Porter et al., *Nucl. Instr. Meth.*, **87**, 1970, 87.

[6] H.E. Bergeson et al., *Phys. Rev. Lett.*, **39**, 1977, 847.

[7] R.M. Baltrusaitis et al., *Nucl. Instr. Meth.*, **A240**, 1985, s410–428.

[8] A.N. Bunner, Cosmic Ray Detection by Atmospheric Fluorescence, Ph.D. thesis, Cornell University, Ithaca, NY, 1967; R.W. Nicholls et al., *Proc. Phys. Soc.*, **74**, 1959, 87; R.H. Hughes et al., *Phys. Rev.*, **123**, 1961, 2084.

[9] R.H. Hughes et al., op. cit., p. 2084.

[10] Ibid., p. 2084.

[11] C.W. Allen, *Astrophysical Quantities*, Athlane Press/University of London, London, U.K., 1976.

[12] B. Rossi, *High Energy Particles*, Prentice-Hall, Englewood Cliffs, NJ, 1952, Chap. 5.

[13] A.M. Hillas, *J. Phys. G: Nucl. Phys.*, **8**, 1982, 1461.

[14] J. Linsley, *Proceedings of the 18th ICRC*, Bangalore, India, 1983, vol. **12**, 135.

[15] P. Halverson and T. Bowen, *Proceedings of the 19th ICRC*, La Jolla, USA, 1985, vol. **7**, 280–283.

[16] J. Linsley, *Workshop on Very High Energy Cosmic Ray Interactions*, Philadelphia, PA, 1982, 476.

[17] F. Arqueros et al., *New J. Phys.*, **11**, 2009, 065011.

[18] P. Sokolsky and G.B. Thomson, *J. Phys. G: Nucl. Particle Phys.*, **34**(11), 2007.

[19] J. Abraham et al., *Nucl. Instrument Methods Phys. Res. Section A*, **523**(1–2), 2004, 50.

[20] H. Tokuno et al., *Nucl. Instrument Methods Phys. Res. A*, **676**, 2012, 54.

6

EXPERIMENTAL TECHNIQUES
Hybrid Detectors

6.1 Introduction

Detectors of ultrahigh energy cosmic rays (UHECR) need to cover very large areas since the flux is so small (1 event per km^2 per century at 10^{20} eV). This leads them to be located in areas of the globe where their presence does not impinge on other human activity, and other human activity will not impinge on them. If a desert location is chosen (or at least a location that is very dry), then one can also construct a fluorescence detector (FD), since water vapor absorbs the near-ultraviolet fluorescence light. Another requirement is very dark skies, which many deserts satisfy. If there is a large plain or valley, covering thousands of square kilometers, one can also build a surface detector (SD). This co-located combination of FD and SD is called a "UHECR hybrid detector." There are two such experiments in operation as of this writing: the Pierre Auger Observatory (PAO) and the Telescope Array (TA) experiment. PAO is located near the town of Malargue, on the Pampa Amarilla in western Argentina, in a cattle ranching area just east of the Andes Mountains. TA is near the town of Delta, in the west desert of Utah, USA, which is located in a valley about 80 × 160 km^2 in area. The SD and FD detectors of Auger and TA were described in previous chapters.

Each of the two detector systems has advantages for studying UHECR: hybrid observation, where the same event is seen by both detectors, combines some of the advantages of both. In particular, hybrid observation typically improves the geometric reconstruction of FD UHECR events by almost an order of magnitude. This improvement goes directly into better energy resolution. More importantly, the resolution in the depth of shower maximum, X_{max}, is greatly improved. In a situation where the difference in X_{max} between proton-induced and iron-induced showers is 80–100 g/cm², an improvement in resolution from 50 (monocular) to 20 (hybrid) g/cm² is very significant. Also, importantly, the better resolution also reduces any systematic bias that may exist in X_{max} reconstruction.

The first hybrid observation of UHECR was performed by joining prototype detectors of the original Fly's Eye experiment to the Volcano Ranch SD experiment,[1] located in New Mexico, USA. The purpose of this pioneering effort was not to run for the long time required to collect enough events for UHECR physics purposes but, rather, to prove that the fluorescence technique worked as expected. The demonstration was successful and the Fly's Eye experiment[2] was subsequently built on the U.S. Army Dugway Proving Ground in Utah. The Casa-MIA SD[3] experiment was built three km from the hill atop which Fly's Eye telescopes were located. This array operated independently, but some events were observed simultaneously in both. The successor to the Fly's Eye was the High Resolution Fly's Eye (HiRes) experiment. When HiRes telescopes were in the prototype stage, the HiRes and Casa-MIA collaborations decided to join together to run what became the HiRes-MIA experiment. This collaboration performed the first physics measurements in hybrid mode.[4]

6.2 Analysis of Hybrid Events

As seen earlier, geometrical reconstruction of an FD event is made by fitting the time vs. angle plot to a tangent function in ψ (the angle in the shower-detector plane). The tangent is sensitive to curvature in this plot, and events with long tracks in the FD camera tend to have more curvature and, hence, better resolution. Since there are strong correlations between ψ and R_p (the perpendicular distance between the FD telescope and the shower track), more curvature also improves the knowledge of distance to the shower. Adding the SD counters' time and angle values to the plot extends the track length, and adds curvature to the plot. Where the ψ resolution for monocular events is about 5°, in hybrid mode the resolution is reduced to less than 1°. The resolution in R_p is correspondingly improved.

Since multiple SD counters are typically struck by an UHECR shower, the question arises as to which counters to use in the hybrid analysis. The Auger hybrid analysis uses a single counter, which is closest to the projection to the ground of the shower-detector plane. This provides most of the improvement in resolution. The TA analysis uses all the struck counters, which gives only slightly better results.

Figure 6.1 shows a hybrid event from the TA data. The four parts of the figure (clockwise from the lower left) show the struck tubes of the

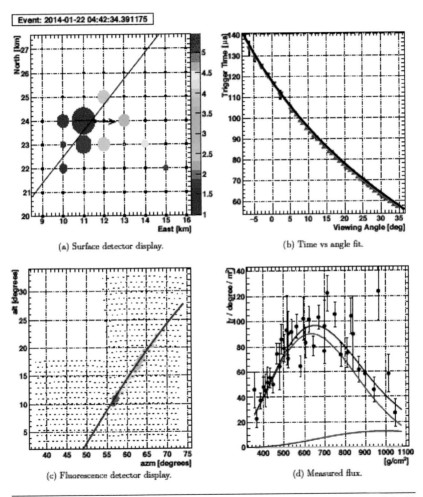

(a) Surface detector display.

(b) Time vs angle fit.

(c) Fluorescence detector display.

(d) Measured flux.

Figure 6.1 A hybrid event from the TA data. The four panels show the struck tubes of the FD (lower left-hand panel), the struck counters of the SD (upper left-hand panel), the time vs. angle plot (upper right-hand panel), and the profile of the energy deposit as a function of atmospheric depth. In the upper right-hand panel, the FD tubes' times are shown in red and the SD counters in blue. Figure courtesy of D. Ivanov.

FD, the struck counters of the SD, the time vs. angle plot, and the profile of FD tubes as a function of shower depth. The upper right figure, where the FD tubes are plotted in red and the SD counters in blue, demonstrates how the time range is extended for hybrid events and the curvature increased.

6.3 Energy Range Covered by Hybrid Events

As seen in Chapter 4, SDs have an efficiency that is energy dependent, up to the knee of the efficiency curve, and at higher energies their efficiency is 100%. Therefore, above the efficiency knee all events whose shower cores are contained will be detected by the SD, and only the FD efficiency has to be calculated. This process is straightforward and discussed in Chapters 3 and 5. At lower energies, the efficiency of the SD has to be calculated as well. At this point, only the TA collaboration has done this. As a result, Auger can measure the spectrum in hybrid mode above $10^{18.4}$ eV, while for TA the minimum energy is $10^{18.2}$ eV.

There is another method to lower the minimum energy which involves a "hybrid trigger." In this case, when an FD trigger occurs, a trigger is forced on the SD in order to catch any counters that may have been struck by the shower. In practice, the trigger rate of an FD is too high to use all triggers, so one imposes further cuts on the FD events (such as track length, or number of struck phototubes) before energizing the hybrid trigger. One can also send the trigger the other way: when the SD trigger fires, it forces an FD trigger. In this way, events below 10^{18} eV can be collected in hybrid mode.

References

[1] H.E. Bergeson et al., *Phys. Rev. Lett.*, **39**, 1977, p. 847.
[2] H.E. Bergeson, J.C. Boone, and G.L. Cassiday, *Proceedings of the 14th International Cosmic Ray Conference*, 1975, Vol. **8**, p. 3059.
[3] A. Borione et al., *Nucl. Instrum. Meth. A*, **346**, 1994, p. 329.
[4] T. Abu-Zayyad et al., *Astrophys. J.*, **557**, 2001, p. 686.

7

EXPERIMENTAL TECHNIQUES
Radio Detection of Cosmic Ray Cascades

7.1 Introduction

Radio techniques as a method for studying cosmic ray air showers are growing in popularity. Extensive air showers (EAS) mainly generate radio pulses by two mechanisms: the geomagnetic effect and the Askaryan effect. Experiments since the early 2000s, together with theoretical efforts, have made considerable progress in understanding how to interpret radio signals to determine characteristics of the cosmic rays that initiate the showers. The frequency band between 30 and 80 MHz is the most well studied. Quantities such as energy, the direction of the source, and even X_{max}, the depth of shower maximum, are accessible, and have been measured by such experiments.

One of the main virtues of radio detectors is that they have almost a 100% duty cycle. This is to be compared with fluorescence detector (FDs), the only other detector systems to measure X_{max} directly, which operate only at night when the moon is down (~10% of the time). In addition, there is almost no atmospheric scattering of radio waves, whereas atmospheric effects are important for the ultraviolet light seen by an FD. The radio pulse is generated by the electromagnetic component of the shower, avoiding many of the uncertainties in Monte Carlo simulations. And, like an FD, the radio signal is calorimetric.

Typically, an array of detectors—each including a radio antenna, associated electronics, power source, and readout system—is deployed on a grid the spacing of which is determined by the footprint of the radio beam generated by the air shower. The beam is narrow, a few degrees wide, and covers about 300 m in diameter on the ground at normal incidence. In order for several detectors to observe the event, spacings of 100–200 m are required. If, for example, the spacing were

150 m, there would be 50 counters per square km. This limits the size of an array: in order to cover 1000 km², which is the area required to collect sufficient statistics for studies of ultrahigh energy cosmic rays (UHECR), 50,000 detectors would be required. While the detectors are less expensive than a typical scintillator or water tank surface detector (SD), the cost of a 1000 km² radio array would be prohibitive— not only in initial investment, but also in upkeep and infrastructure costs. The largest array built to date is the Auger Engineering Radio Array (AERA)[1] of 153 detectors.

Radio noise backgrounds are a problem. Most experiments have been co-located with a cosmic ray detector of another sort, such as a ground array, hopefully at a location of low radio background noise. Even in low-noise environments, self-triggering of the radio experiment has been found to be difficult—exceptions include the TREND[2] experiment located in rural China, and the Askaryan Radio Array (ARA)[3], the Antarctic Impulsive Transient Antenna (ANITA),[4] and the Antarctic Ross Ice-Shelf ANtenna Neutrino Array (ARIANNA)[5] experiments in Antarctica—so, the existing cosmic ray detector is used for triggering, as well as for comparison of experimental results.

The energy range covered by existing arrays is determined by two factors. The low energy limit comes from the signal size compared to the background noise level; the noise minimum is that of the Milky Way galaxy and is typically about 10^{17} eV. The high energy limit comes from the dropping flux of cosmic rays with energy. A 10 km² array, for example, can only reach up to about 10^{18} eV. This energy range seems limited but is actually the energy range of the galactic–extragalactic transition (see Chapter 9) and is thus very interesting.

7.2 The Geomagnetic Effect

This is the main mechanism for the generation of radio pulses by a cosmic ray air shower. As particles in a cosmic ray air shower propagate down through the atmosphere, they are subject to the Lorentz force of the Earth's magnetic field. Electrons are pushed in one direction and positrons in the opposite direction. However, this separation of charge by itself does not generate a radio wave, since the frequency

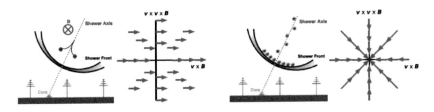

Figure 7.1 Polarization of geomagnetic and Askaryan radiation. The left-hand panel shows that geomagnetic radiation is polarized along the $\mathbf{v} \times \mathbf{B}$ direction. The right-hand panel shows that the Askaryan radiation is polarized in a radially inward direction. From T. Huege, *Phys. Rep.*, 620, 2016, 1, with permission.

of interactions between electrons and air molecules is very high. Just as electrons in a conductor drift subject to a voltage difference, so the shower electrons and positrons drift apart, transversely to the shower direction, at an average speed. One could call these "transverse currents." There is, however, an additional effect: the number of shower particles changes as the shower develops, reaching a shower maximum, and then waning. In effect, this produces a time variation of the transverse currents which, in turn, generates a radio pulse. The radio signal is polarized in the direction of the Lorentz force, $\mathbf{v} \times \mathbf{B}$. This polarization is indicated on the left-hand side of Figure 7.1.

Since it is coming from an almost planar shower front, the pulse is a narrow beam, only a few degrees across, and illuminates about 0.1 km² on the ground.

7.3 The Askaryan Effect

A second, but weaker effect that generates a short radio pulse is the Askaryan effect. Here, knock-on electrons from air molecules struck by shower particles add up to a 10–20% net negative charge excess for the electromagnetic component of the shower. Once again, it is the change in shower particle numbers as the shower waxes and wanes that generates the radio pulse. This contribution is independent of the geomagnetic field and occurs even if the shower direction is parallel to the field.[6]

The polarization of the Askaryan radio wave is radially inward, as shown in the right-hand side of Figure 7.1. For an array of radio detectors observing a shower, there are individual detectors where the polarizations of the geomagnetic and Askaryan contributions are

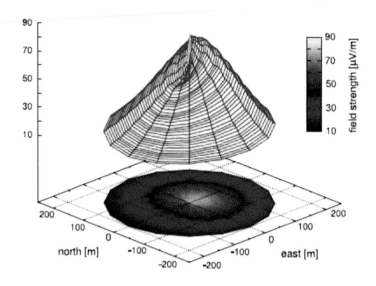

Figure 7.2 Simulation of an event at the location of the LOPES experiment. The addition of the radio waves from the geomagnetic and Askaryan effects produces an asymmetric signal in the 40–80 MHz band. From T. Huege, *Phys. Rep.*, 620, 2016, 1, with permission.

parallel and detectors where they are in opposition. This leads the radio signal to be asymmetric as a function of location around the shower axis. Figure 7.2 shows this for a simulation at the location of the LOPES experiment.[6] Again, the footprint of the radio pulse on the ground is about 0.1 km².

An additional interesting and important effect is that the signals from different parts of the shower are approximately coherent, since the propagation of the radio waves and the shower front both occur at the speed of light. The coherence is larger at lower frequencies. The electric fields from the whole shower add linearly, and are approximately proportional to the total number of electrons and positrons in the shower and, hence, to the energy of the cosmic ray. The power in the radio beam is quadratically related to the cosmic ray energy.

7.4 Reconstructing Cosmic Ray Properties from Radio Measurements

The three most important properties of cosmic rays that a radio experiment seeks to reconstruct from the radio pulses are the direction of the source, the energy, and the depth of shower maximum.

To reconstruct the direction of the source, the timing of hit counters is utilized in a similar way as that described in the SD analysis in Chapter 4. To determine the core position, the full asymmetry of the shower intensity (due to positive and negative interference between the geomagnetic and Askaryan signals) must be taken into account.

To determine the cosmic ray's energy, the basic reconstruction starts with a method of handling the asymmetry of the shower intensity. One method is, first, to symmetrize the pulses from different detectors, taking the interference into account; and, then, to measure the lateral distribution of counter pulse heights. This leads to having an energy estimator (often the radio intensity 50 m from the core) that is related to shower energy, either by Monte Carlo simulations (this is more model-dependent than one would like), or by comparison with the result of a co-located cosmic ray detector of another sort.

Alternatively, the full asymmetric distribution of antenna pulses can be used; for example, by comparison with a look-up table of distribution shapes. Then, the absolute intensities of the signals can be used to scale the energy from the look-up table value. Comparison of measured energies with a co-located cosmic ray detector is always important.

Two further advantages of radio detection are that it is generated by the electromagnetic component of an air shower, which is the most well-understood component; and that the signals are coherent over the length of the shower. Thus, the radio intensity is proportional to the square of the cosmic ray's energy.

The energy threshold is determined by the requirement to have signals in ~3 or more detectors for reconstruction accuracy. The detector spacing limits this for small zenith angles. Although events at large zenith angles have an elongated footprint on the ground, and strike more detectors, each signal is smaller, which raises the energy threshold. The highest energy that can be used for physics measurements is set by the size of the radio array, the running time of the detector, and the flux of cosmic rays. As of the time of writing, the Tunka–Rex[7] collaboration, for example, has shown a spectrum result covering the energy range $17.2 < \log(E/eV) < 18.0$.

The aperture of the experiment as a function of energy must be calculated using Monte Carlo techniques. Shower simulation programs such as CORSIKA have been married to radio generation routines (two such programs are CoREAS[8] and ZHArieS[9]), producing an accurate result. Since cosmic rays arriving approximately parallel to the geomagnetic field at the experiment's location have much smaller pulse heights, there is no energy above which the experiment is 100% efficient. However, one can cut out an area in zenith angle–azimuthal angle space where the inefficiency lies, resulting in a constant aperture for the remaining space. For anisotropy studies, this creates a slice in declination on the sky of lower efficiency, which is unfortunate.

For X_{max} reconstruction, what is actually measured is the distance from the ground to the X_{max} point of the shower. Using the density distribution of the atmosphere, X_{max} can then be calculated in g/cm^2 units. This distance is determined by the "flatness" of the distribution of detector pulse heights. More curvature means a smaller distance to X_{max}. X_{max} reconstruction of very deep showers is poor because of the short distance from detector to X_{max}. For events with X_{max} high in the atmosphere, or for high zenith angle events, the resolution is poor as, at high distances, all showers look the same.

7.5 Experiments

Figure 7.3 shows, on the same distance scale, the layout of several radio experiments. The aim of these experiments was primarily to prove that radio techniques are valid for the measurement of cosmic ray properties. Experiments in Antarctica and the TREND experiment are not shown; neither is The AugerPrime[10] radio experiment. TREND is located in a very rural location in XinJiang, China, and is important for two reasons: it has been successful in self-triggering, and it is a pre-prototype for the GRAND experiment. GRAND is a very ambitious experiment to deploy tens of thousands of radio antennas to search for tau neutrinos, and also to make cosmic ray measurements.

As described, measurements of the spectrum of cosmic rays by radio techniques are presently rare, only limited results on X_{max} measurements of selected events (and not on the total distribution) are available, and anisotropy results are nonexistent. While promising, this technique is just now becoming established.

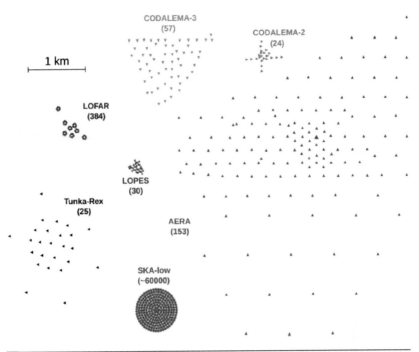

Figure 7.3   Layouts of various radio experiments, all on the same distance scale. The largest is the Auger Engineering Radio Array (AERA), which covers 17 km² in a graded array. Not shown are the TREND and AugerPrime arrays, or experiments located in Antarctica. From T. Huege, *Phys. Rep.*, 620, 2016, 1, with permission.

Tunka-Rex is co-located with an atmospheric Cherenkov array for triggering, and seeks to measure cosmic ray properties. LOFAR[11] is co-located with a small scintillation counter array for triggering. The Square Kilometer Array[12] is a comprehensive astrophysics array in Australia that will start observing in 2020, has a very dense distribution of antennas, and includes measurements of cosmic ray properties in its long list of priorities.

The AERA is located in the western part of the Pierre Auger experiment, which can be seen in Figure 7.3 to be a graded array. Since the background noise level is high, AERA is triggered by other detectors in the Auger suite. Experience with AERA has led the Auger collaboration to include a radio component in their upgrade program, called AugerPrime.

The aim of AugerPrime is to study anisotropy by choosing which air showers are initiated by protons and light nuclei (and thus have been bent to a lesser degree by extragalactic and galactic magnetic fields). The method chosen for executing this is to make simultaneous measurements of the muonic and electromagnetic components of every individual event. Events with the lowest muonic/electromagnetic component ratios would presumably be those to choose for anisotropy analysis. The Auger SD water tanks will provide the muon measurements. For events below zenith angles of about 60°, scintillation counters to be added atop the water tanks will provide electromagnetic component measurements. For highly inclined showers, the electromagnetic component does not reach the ground, but the radio signal does. Hence, the inclusion of radio antennas, also atop the water tanks, that will be sensitive to the size of the electromagnetic component of air showers. Triggering for both scintillation counters and radio antennas will come from the water tank detectors. Here, the experimenters' intent is not to use radio for measurements of spectrum, X_{max}, or anisotropy but, rather, only for a relative measurement of the size of an air showers' electromagnetic component.

7.6 Summary

The development of radio techniques since the turn of the millennium has been impressive. Experiments have mostly been dedicated to understanding the radio technique, and comparisons of events using Monte Carlo simulations (which have been developed simultaneously) have resulted in good agreement, indicating that this understanding has been accomplished. Measurements of the direction of air showers, their energy, and the depth of shower maximum have been made. Experiments to extend this engineering information to actual cosmic ray physics have begun and more are planned.

Except in the most rural of locations, radio experiments have not been able to self-trigger. Since, for engineering purposes, one wants to compare results of a new technology to those of an accepted technology, co-location with other cosmic ray detectors has been the rule. This is likely to continue to be the case. Even in these locations, the task of understanding whole experiments, rather than just individual events, using the Monte Carlo technique is just beginning.

One aspect of the radio technique, searching for ultrahigh energy neutrinos in Antarctica, has not been mentioned here. The ANITA balloon-borne experiment has been flown several times on long duration balloons that circle the South Pole. Cosmic rays have been seen (but not neutrinos), and energies estimated where fluxes are in the right ball park to be cosmic rays. The exposure of ANITA is very small, however. The ARA (at the South Pole) and ARIANNA (on the Ross Ice Shelf) experiments are also looking for ultrahigh energy neutrinos, but have a cosmic ray component. These three experiments perform self-triggering successfully.

References

[1] A. Aab et al., *Phys. Rev. D*, **93**, 2016, 122005.
[2] D. Ardouin et al., *Astropart. Phys.*, **34**, 2011, 717.
[3] P. Allison et al., *Astropart. Phys.,* **35**, 2012, 457.
[4] S.W. Barwick et al., *Phys. Rev. Lett.,* **96**, 2006, 171101.
[5] S.W. Barwick et al., *Astropart. Phys.,* **70**, 2015, p. 12.
[6] T. Huege, *Phys. Rep.,* **620**, 2016, 1.
[7] P.A. Bezyazeekov et al., *Nucl. Instrum. Meth. A,* **802**, 2015, 89.
[8] T. Huege, M. Ludwig, C.W. James, *AIP Conf. Proc.,* **1535**, 2013, p. 128.
[9] J. Alvarez-Muniz et al., *Astropart. Phys.,* **35**, 2012, p. 287.
[10] C. DiGiulio et al., *Nucl. Part. Phys. Proc.,* **279**, 2016, p. 153.
[11] P. Schellart et al., *Astron. Astrophys.,* **560**, 2013, A98.
[12] T. Huege et al., *Proceedings of the Science PoS (ICRC2015),* 2015, 309.

8

THE COSMIC RAY SPECTRUM

8.1 Introduction

Measurements of the spectrum of cosmic rays show that it extends from MeV energies to above 100 EeV. The highest energy event,[1] at 320 EeV, was seen by the Fly's Eye experiment in 1991. The process of measuring the spectrum consists of detecting events, determining their energy, and calculating the efficiency of the detector. The flux, $J(E)$, is given by Equation (8.1):

$$J(E) = N(E)/(A\Omega T \in (E)\Delta E) \tag{8.1}$$

where $N(E)$ is the number of events in the energy bin of energy E, A is the area covered by the detector, Ω is the solid angle covered by the detector, T is the observation time, $\in(E)$ is the efficiency of the detector at energy E, and ΔE is the width of the energy bin at E. In other words, calculating the flux takes out the specific details of the experiment, such as area and solid angle, so all experiments should obtain the same answer.

The overall picture of the spectrum is that the flux falls steeply with energy, following a power law of index -2.7 with very little variation, up to a feature known as the "knee" in the middle of the 10^{15} eV decade, above which it steepens to a nominal $E^{-3.1}$ power. The spectrum approximately follows this power law up to the highest energies (see Figure 8.1).

A power law is predicted if the Fermi acceleration mechanism operates in the sources of the cosmic rays. In this case, a moving magnetic field intercepts particles in its path and accelerates them, the upper limit of the spectrum being determined by the strength of the field and the time or distance through which it is in contact with the particles. An example of this picture is a galactic supernova remnant: the outer layers of the star are ejected in the explosion and carry a magnetic field

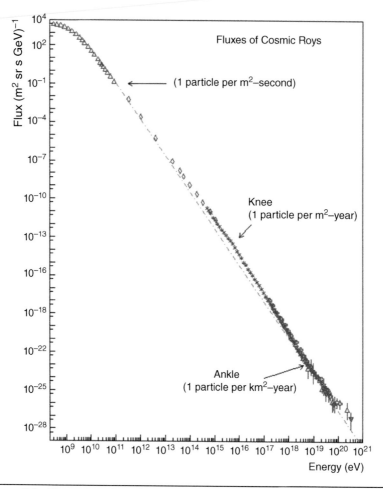

Figure 8.1   The spectrum of cosmic rays from GeV to hundreds of EeV energies. The flux follows an $E^{-2.7}$ power law up to the knee, then steepens to about $E^{-3.1}$. Other spectral features, such as the ankle, occur above the knee.[2]

with them as they expand to form the remnant. Predictions of the upper energy limit are in the 10^{14} eV decade, and mechanisms for reaching the knee of the spectrum have been proposed. Therefore, from the point of view of potential sources, the lower energy portion of the spectrum seems consistent with a galactic origin.

In the case of an expanding magnetic field, the maximum energy of a source is proportional to the atomic number of the nucleus being accelerated. In this picture, we can predict what the end of the galactic part of the spectrum would be. The knee would be at the maximum energy of protons, while helium reaches its maximum at a factor of 2 higher energy, and so on. Since Li, B, and Be are rarer, one expects there is likely a dip in the spectrum where their maxima would be, after which the spectrum would rise again because of the higher abundances of C, N, and O. The galactic spectrum would start to come to an end a factor of 26 above the knee because Fe is the heaviest element that can be created exothermically in nuclear fusion. We will see that this expectation is borne out in the data.

Above the knee, there must be a transition to extragalactic sources since, at the highest energies, the galactic magnetic field is not strong enough to deflect particles significantly and there would be a strong anisotropy pointing back to galactic sources that is not observed. Hence, the highest energy events must be of extragalactic origin. Many galaxies, perhaps all, have a supermassive black hole at their centers. The Milky Way has one, of several million solar masses, which is not bright at any wavelength. However, many black holes are very active, and several mechanisms have been proposed allowing them to accelerate cosmic rays to the highest energies. Again, the magnitude of an expanding magnetic field and the size of the accelerating region (which is equivalent to the time the field and particles are in contact) determine the maximum energy reached by the source. This is shown in the Hillas plot (Figure 8.2).

The shape of the extragalactic cosmic ray spectrum is modified by the propagation from source to Earth. Inelastic pion production interactions between cosmic ray protons and the cosmic microwave background radiation (CMBR), and photons of other wavelengths, create what is called the "GZK cutoff" at about 60 EeV after about 50 megaparsecs (Mpc) of propagation.[4] For sources farther away, propagation effects (e^+e^- pair production in these same collisions) create a dip in the spectrum at about 5 EeV called the "ankle."[5] For cosmic rays that are atomic nuclei, spallation by the photon fields knocks out approximately 1 nucleon for every Mpc of travel. At the highest energies, this limits the distance nuclei can propagate to 50 Mpc for iron (and creates a cutoff at about 60 EeV) and 1–2 Mpc for helium, with other elements in between. We can thus expect that there will be structure in the spectrum of extragalactic cosmic rays, too.

Because of the arguments given above, the spectrum from the knee to the highest energies should be different from the spectrum at lower

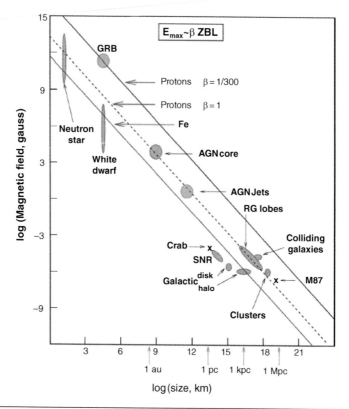

Figure 8.2   A version of the Hillas plot showing the relation of magnetic field magnitude and size of the accelerating volume at 100 EeV. Estimates of the location of various proposed source types are plotted also.[3]

energies. Structures should appear that we can interpret to gain considerable information about the sources of cosmic rays and the distances cosmic rays travel from source to Earth.

8.2 The End of the Galactic Cosmic Ray Spectrum

Figure 8.3 shows the combined spectrum measured using the TA surface detector (SD), and TA Low Energy Extension (TALE) fluorescence detector (FD)[6] in monocular mode. The TALE spectrum, shown in red, extends from $10^{15.3}$ eV to $10^{18.3}$ eV, and exhibits 3 features, called the knee, the dip at about $10^{16.2}$ eV, and the second knee at about $10^{17.1}$ eV. At $10^{18.2}$ eV, the TA SD spectrum, shown in blue, starts up, and continues to $10^{20.3}$ eV. Two further features are seen, the ankle at $10^{18.7}$ eV and the cutoff at $10^{19.8}$ eV.

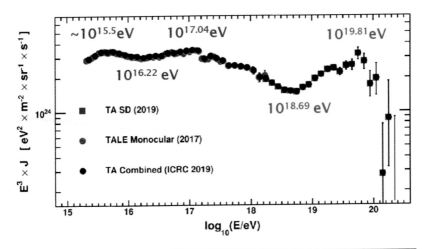

Figure 8.3 The TA and TALE combined spectra, multiplied by E^3 to emphasize the features. The spectra cover five orders of magnitude in energy, and five spectral features are to be seen: the knee at $10^{15.6}$ eV, the dip at $10^{16.2}$ eV, the second knee at $10^{17.04}$ eV, the ankle at $10^{18.7}$ eV, and the cutoff at $10^{19.8}$ eV. From D. Ivanov et al., *Proceedings of Science PoS (ICRC2019)*, 2019, 298, with permission.

The irregularity in the TALE spectrum at $10^{17.25}$ eV is not borne out by further TALE data, so we will ignore it while we interpret the remainder of the TALE spectrum. An interpretation of this set of spectral features is that we are seeing a "rigidity-dependent sequence" of elemental cutoffs. Rigidity is the energy/charge of cosmic ray nuclei, and the hallmark of the Fermi mechanism of cosmic ray acceleration is that it accelerates all nuclear species up to a constant rigidity. The interpretation given in the following paragraph indicates that galactic sources reach about 4 PV in rigidity.

The knee is about a factor of 26 lower than the second knee. Because it is a sharp feature, let us start by interpreting the second knee as the "iron knee;" that is, it is located at the maximum energy reached by galactic sources accelerating iron. This implies the energy of the proton knee should be $10^{15.6}$ eV, with helium located at about $10^{15.9}$ eV. The broad structure seen in the TALE spectrum seems to be both the proton and helium features. Lower energy direct measurements indicate that the flux of protons and helium are very similar, and this is borne out in the TALE measurement. A beryllium knee should appear at $10^{16.2}$ eV, which is the center of the spectral dip. This can be accounted for by the relative rarity of Li, Be, and B. Carbon should appear at $10^{16.4}$ eV, where the spectrum is rising. Other elements also contribute to the rise to the iron knee.

This interpretation of the 10^{15}–10^{17} eV energy range has implications for cosmic ray composition measurements, which we will address in a later chapter.

8.3 The Galactic–Extragalactic Transition and Extragalactic Spectrum

Figure 8.4, at the top, shows the spectrum measurements reported at the 2019 International Cosmic Ray Conference by TA and Auger,

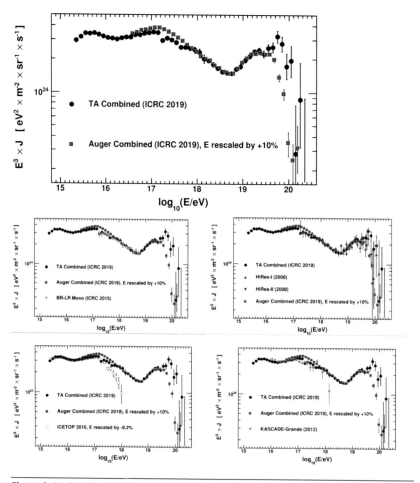

Figure 8.4 Top: TA and TALE combined spectrum compared with the Auger combined spectrum (with energy scale raised by 10%). Center left: TA and Auger spectra plus the TA FD monocular spectrum. Center right: TA and Auger plus the HiRes monocular spectra. Lower left: TA and Auger spectra plus the IceTop spectrum (energy scale lowered by 9%). Lower right: TA and Auger spectra plus KASCADE-Grande spectrum. From D. Ivanov et al., *Proceedings of Science PoS (ICRC2019)*, 2019, 298, with permission.

with the Auger energy scale raised by 10% (within the reported systematic uncertainties of both experiments) to achieve agreement in the ankle region ($10^{17.8}$ eV to $10^{19.1}$ eV). The difference between the results of the two experiments in the energy of the cutoff stands out (Auger: $10^{19.6}$ eV; TA: $10^{19.8}$ eV). Small differences also exist at lower energies ($10^{16.5}$–$10^{17.8}$ eV).

Also seen at the top of Figure 8.4 is a new feature emerging in the Auger spectrum: a shoulder at $10^{19.1}$ eV. There is a hint of this feature in the TA SD spectrum, too. It is interesting to note that this energy is the threshold for helium nuclei being broken up by spallation by the CMBR.

The center and lower panels of Figure 8.4 show both the TA and Auger combined spectra plus four other spectrum measurements, by TA FD monocular, HiRes monocular, IceTop (energy scale lowered by 9%), and KASCADE-Grande. All four agree with the TA/TALE combined spectra. There is therefore a consensus that there are three features below 10^{18} eV, and two (with a third now emerging) above that energy.

Figure 8.5 is taken from a comparison by the TA collaboration of their SD data in the lower half of the 10^{18} eV decade to models of what galactic and extragalactic sources would produce.[7] The conclusion that emerged is that, at the 95% confidence level, fewer than 1.3% of events in this energy range are of galactic origin. Although this result seems model-dependent, the robust result is that the vast majority of events in this energy range are of extragalactic origin. Any galactic magnetic field model with the regular component parallel to the galactic arms would look like this. Above, we have interpreted the three spectral features below 10^{18} eV as a rigidity-dependent cutoff at the high end of the galactic cosmic ray spectrum. Figure 8.5 indicates that the galactic–extragalactic transition is complete by 10^{18} eV. This leads us to the conclusion that the features above this energy are imprinted on the spectrum by extragalactic sources and propagation from source to Earth.

There are several mechanisms that could create the high energy cutoff at about 60 EeV. First, the cutoff could be the true GZK limit, which is correct only for a mostly protonic composition. This mechanism, combined with a protonic flux, also predicts that the ankle must exist, it must be at the energy at which it is seen, and must have the degree of inflection shown in the data. This scenario also predicts that there should be a flux of neutrinos from the GZK process;

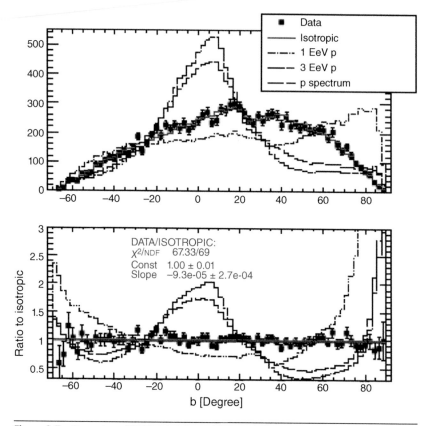

Figure 8.5 Comparison of TA SD data between $10^{18.0}$ and $10^{18.5}$ eV as a function of galactic latitude, b, with predictions based on galactic magnetic field models. Black points are the data, the black lines are predictions based on the assumption that the origins of these events are galactic objects, and the red line is the assumption that the origins of events in this energy range are extragalactic. The upper panel is a histogram of events in galactic latitude. The lower panel is the ratio of the data to expectations of an isotropic distribution. From R.U. Abbasi et al., *Astropart. Phys.*, 86, 2017, 21, with permission.

experiments are searching for this, but have not yet found it. A second mechanism could be that the 60 EeV cutoff is the maximum energy of the sources. Here, there is not necessarily an ankle, and the composition need not be protonic. A third scenario is that the composition at the cutoff is heavy (mostly Fe) and spallation by the CMBR causes the cutoff. Here, in general, there is not necessarily an ankle. Therefore, more information is needed to resolve this question, from UHECR composition measurements and from neutrino measurements.

The difference in ultrahigh energy spectra, particularly in cutoff energies, between TA and Auger has prompted the two collaborations

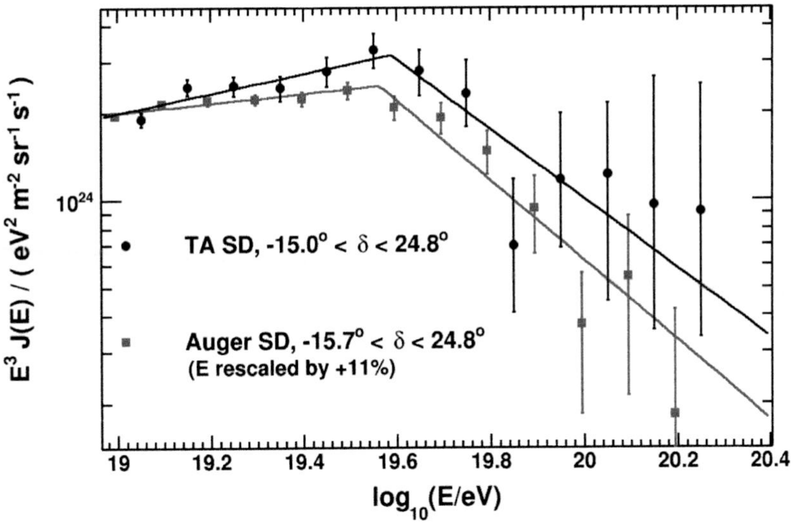

Figure 8.6 TA and Auger spectra measured in the common declination band. The cutoff energies are in good agreement, at $10^{19.57}$ eV. In this figure, the TA energy scale has been lowered by 5.2%, and the Auger energy scale raised by 5.2%, so as to achieve agreement in the ankle region. The mismatch (Auger lower, TA higher) shown here must be due to a further effect. From D. Ivanov et al., *Proceedings of Science PoS (ICRC2019)*, 2019, 298, with permission.

to search jointly for their origin. The experimenters concentrated on measuring the spectra of the two detectors in the part of the sky where their coverages overlap.[8] This is from about −15 to +25° in declination. Figure 8.6 shows the result of this search: in the common declination band, the two cutoff energies agree well. This indicates that there is likely a difference in cutoff energies in the northern sky, with the cutoff energy being higher at higher declinations. This difference will be discussed in Chapter 9.

In Figure 8.6, the TA and Auger energy scales have been modified to achieve agreement in the ankle region. The fact that the Auger spectrum is lower than the TA spectrum in Figure 8.6 indicates that there is a further energy-dependent effect in play.

References

[1] D.J. Bird et al., *Astrophys. J.*, **441**, 1995, 144.
[2] M. Boratav, *Europhysics News*, **33**, 2002, 162.
[3] A. Letessier-Salvon and T. Stanev, *Rev. Mod. Phys.*, **83**, 2011, 907.

[4] K. Greisen, *Phys. Rev. Lett.,* **16**, 1966, 748; G.T. Zatsepin and V.A. Kuz'min, *JETP Lett.*, **4**, 1966, 78.

[5] R. Aloisio et al., *Astrophys. J.*, **27**, 2007, 76.

[6] D. Ivanov et al., *Proceedings of Science PoS(ICRC2019)*, 2019, 298.

[7] R.U. Abbasi et al., *Astropart. Phys.*, **86**, 2017, 21.

[8] O. Deligny et al., *Proceedings of Science PoS(ICRC2019)*, 2019, 234.

9

SEARCHES FOR ANISOTROPY

9.1 Introduction

Although we have considerable knowledge about cosmic rays, not only those of galactic origin, but also those of extragalactic origin (see Chapter 8 on spectrum, and Chapters 10, 11, and 12 on composition), we do not know their sources. We cannot point to any astronomical object, or even any class of astronomical objects, and say, "This is a source of cosmic rays." There are two reasons for this. First, cosmic rays do not point back toward their sources. Galactic and extragalactic magnetic fields bend their trajectories by angles that are poorly known for galactic events, or unknown in the case of extragalactic events. In the latter case, the highest energy events, if protons, may point away from their sources by 5–20°.

The second reason for the lack of knowledge regarding sources of cosmic rays is that the statistics are too poor to come to any definite conclusions. This is true even at the highest energies, where magnetic fields have the smallest effect. The steeply falling spectrum makes these events so rare that even the largest experiments have difficulty amassing enough events to make a difference. In other words, the cosmic ray sky looks quite isotropic at all energies.

The result of this is that cosmic ray physicists are reduced to searching for any anisotropy at all, even if it is not interpretable as the result of a point source. In this chapter, we describe the state of the art in anisotropy searches.

There seem to be two areas of the sky, each about 40° in diameter, where an excess of events is seen, one in the northern hemisphere known as the TA hotspot and one in the south known as the Auger warm spot. Both have a statistical significance of about 3 standard deviations. The Auger experiment has also found a dipole on the sky that is of 5.2 sigma significance,[1] but the interpretation of a dipole in terms of cosmic ray sources is difficult. We are guided by the rule for publications in *Physical Review Letters*: to say that one has "observed" a new effect, one must be at the 5 sigma or higher level, so that the Auger dipole is an observation of anisotropy.

Two other searches that we mention here are for the variation of the energy of the cutoff in a high declination part of the sky compared to the TA–Auger common declination band (the difference is of 4.3 standard deviation significance), and the correlations in energy and the angle of high energy events that could be caused by extragalactic magnetic fields. This is of about 4.1 sigma significance. In the latter search, one picks a location on the sky and looks for events nearby where, the further away from the chosen point an event is, the lower its energy; that is, there is an anticorrelation between energy and position on the sky. From the strength of the anticorrelation, one can estimate the integral of the field along the path taken by the cosmic rays. The first estimate is 15 nanoGauss megaparsecs (Mpc), which is a reasonable value.

It is interesting to consider what a sky map would look like if we had infinite statistics at the highest energies. It would be bumpy, with the closest and brightest sources showing up as blobs of radius 5–20°. This size is caused by the irregular component of the extragalactic magnetic field. The blob centers would point several degrees away from the source because of the net magnetic field. Therefore, the blob centers would correlate poorly with known source types. Almost all the techniques in use today would have a hard time finding the sources. But there are two things on the horizon that would help greatly. First, the energy–angle correlation method would be able to identify the sources as the positions around which the correlations occur. Second, there are several projects astronomers are starting to undertake to better measure the galactic magnetic field, and to begin measuring the extragalactic field.

9.2 TA Hotspot and Auger Warm Spot

The simplest method of searching for anisotropy is to place a dot on a sky map from which each event points. Figure 9.1(a) is an example of

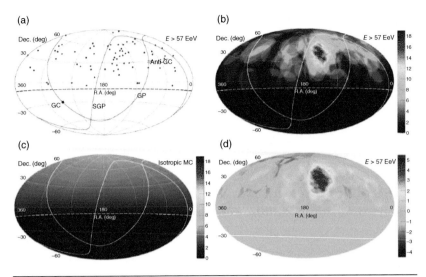

Figure 9.1 Different views of the TA hotspot. Panel (a) shows the events of energy above 57 EeV plotted in equatorial coordinates; panel (b) shows the result of oversampling with a radius of 20°; panel (c) shows the expected sky map if events were exactly isotropic; panel (d) shows the local statistical significance calculated from the Li-Ma method. From R.U. Abbasi et al., *Ap. J.*, 790, 2014, p. L21, with permission.

such a map, in equatorial coordinates, from the 2014 TA paper about the hotspot.[2] The zero of right ascension is on the right-hand end of the projection of the Earth's equator. The map was made from events seen by the TA surface detector above an energy of 57 EeV. It shows a region of higher density centered near RA = 145° and declination = 45°, with a radius of ~20°. Once the physicist's eye identifies such a region, the question becomes how the probability should be evaluated that this over-density occurred by chance. One way is to use oversampling.

An oversampling plot, as shown in Figure 9.1(b), is made by taking each point on the sky as the center of a circle of some radius (in Figure 9.1(b), it is 20°), counting the number of events in the circle, and plotting that number at the point. This is a way of taking into account the size of extended sources. Figure 9.1(c) shows what a completely isotropic sky would look like, and Figure 9.1(d) shows the statistical significance of each point on the sky. The maximum statistical significance of the data (known as the "local significance") is not the answer to the significance question as, in the case that is shown here, it is not centered on an *a priori* location: a chance occurrence could occur anywhere on the sky.

To calculate the global significance, one uses the Monte Carlo method. One throws the number of events in the data in random positions in the experiment's aperture and calculates the maximum local significance in the same way that was used for the data. Then one repeats these trials many times, counting the proportion of the time the local significance equals or exceeds that of the data. One can calculate the number of standard deviations of the data by determining the number of sigmas of a Gaussian function that would give this proportion.

In the 2014 data, and in subsequent data, the global significance of the TA hotspot has varied somewhat but has been near 3 standard deviations. The choice of 57 EeV as a minimum energy comes from previous anisotropy studies that used this minimum energy and is an *a priori* choice.

In applying the oversampling technique to both TA and Auger published data above 57 EeV, the result was Figure 9.2, covering the entire sky, which shows two clusters of events. The Auger warm spot in the southern hemisphere is in a very interesting location, being on the supergalactic plane and almost centered on Centaurus A, the closest active galactic nucleus. Conversely, the TA hotspot in the

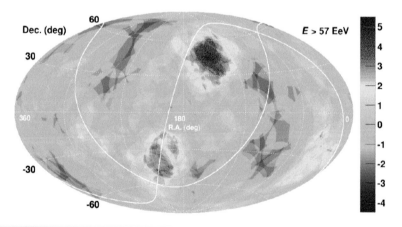

Figure 9.2 Sky map in equatorial coordinates showing the TA hotspot in the northern hemisphere, and the Auger warm spot in the southern hemisphere. This is a significance plot made by oversampling with a 20° radius using published TA and Auger data of energy greater than 57 EeV. The white line rising vertically in the center of the plot is the supergalactic plane, and the U-shaped line is the galactic plane. From K. Kawata, with permission.

northern hemisphere is 19° from both the supergalactic plane and the Ursa Major galaxy cluster. One similarity between the two effects is that, in both locations, there are galaxy filaments in the local large-scale structure.

9.3 The Auger Dipole

Figure 9.3 shows a sky map in equatorial coordinates of Auger SD data oversampled with a radius of 45°. The left-hand panel is for energies between 4 and 8 EeV, and the right-hand panel includes events of energies greater than 8 EeV. One can see that, for the higher energy panel, there is an excess centered near 90° in right ascension and –30° in declination.

To evaluate the significance of the excess, the Auger collaboration performed a Fourier analysis on their data, choosing right ascension and the azimuthal angle of events as their variables. They calculated the dipole and quadrupole amplitudes for a Fourier sine distribution (a_k) and a cosine distribution (b_k), where $k = 1$ (2) is the dipole (quadrupole) component. The total amplitude is r_k, and the probability that an amplitude equal to or larger than r_k arises from an isotropic distribution is $P(\geq r_k) = \exp(-N(r_k)^2/4)$, where N is the number of events in the sample. The probability for the data shown in Figure 9.3 is 2.6×10^{-8}, which is about 5 standard deviations. This is considered an observation of anisotropy.

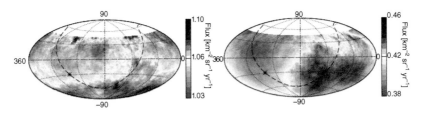

Figure 9.3 Sky maps of Auger SD data oversampled with a radius of 45°. The left-hand panel is for energies from 4 to 8 EeV and the right-hand panel is for energies >8 EeV. From A. Aab et al., *Ap. J.*, 864, 2018, p. 4, with permission.

9.4 Variation of the Cutoff Energy with Declination

In Figure 8.4, we saw that the high energy cutoff occurs at different energies in the TA and Auger spectra but, in the declination range covered by both experiments, the difference disappears. This leads to the two TA flux measurements shown in Figure 9.4. The energy of the cutoff at lower declinations (below 24.8°) is $10^{19.59}$ eV, whereas it is $10^{19.85}$ eV (82% higher) at declinations above 24.8°. To calculate the global significance of the difference, Monte Carlo trials were generated with the same number of events and the same spectrum as the data. The same calculation as was performed on the data was repeated for each trial and the number counted where the local significance was higher than the data. The result was that the global significance is 4.3 standard deviations.

Despite an exhaustive search for an instrumental effect to explain this difference, none was found, and the TA collaboration concluded

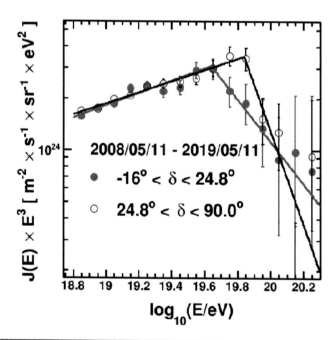

Figure 9.4   TA flux measurements in two different declination ranges. The flux times E^3 is plotted here. The lower declination range, with data in closed circles, is that seen by both TA and Auger experiments. The open circles are measurements at more northerly declinations. From D. Ivanov, with permission.

that it is an astrophysical effect. One possible explanation is that, in the northern sky, there is a source—or perhaps several sources— that are quite close to the Earth and are modifying the observed spectrum.

An interesting question is where the sources are in the sky. The TA collaboration, as yet, has not divided the sky further to perform a more pointed search, citing lack of statistics. In addition, further trials performed on the data mean that the same trials must be performed on Monte Carlo simulations, the result of which is the dilution of the statistical significance of the effect.

9.5 Energy–Angle Correlations

This is a promising method with which to search for anisotropy. The ansatz is that extragalactic magnetic fields along the line of sight between a source and the Earth will bend lower energy cosmic rays to a greater extent than higher energy cosmic rays, producing an anticorrelation between events' energies and angular distances from the source. For example, if there were a single source, higher energy events would be seen to come from a position on the sky closer to the source than lower energy events. Because all cosmic rays are positively charged, if there were a constant magnetic field, the particles would be seen on one side of the source, where the dispersion would string them out in a line perpendicular to the magnetic field. If there were, in addition, regions of magnetic field pointing in random directions, lower energy events would stray from the line to a greater extent than higher energy events. This would produce a triangle-shaped set of events, with one corner on the source, where there would be an anticorrelation between events' energies and positions.

This example suggests how to search for energy–angle correlations. The variables are the orientation of the triangle, its length and width, and the minimum energy of events that would be correlated (low energy events could be thrown over very wide areas and be impossible to find). A search for correlations thus requires a large number of trials, where each trial is a choice of these variables. The left-hand panel of Figure 9.5 shows an example of one of these triangles, and the right-hand panel shows the correlation between $1/E$ and the angle from the triangle apex.

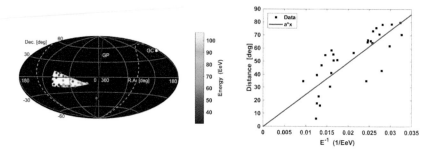

Figure 9.5 Triangle on the sky and the correlation between $1/E$ and angle from the triangle vertex. The left-hand panel shows an example of a triangle that is part of the energy–angle anticorrelation study. The right-hand panel shows, for the events in the triangle, the $1/E$ vs. the angular distance from the triangle vertex. From D. Ivanov, with permission.

With the existing statistics above an energy of, for example, 10 EeV, finding individual sources is very unlikely. However, the Virgo supercluster, of which the Milky Way galaxy forms a part, stretches across the sky in a great circle and includes most of the galaxies within the GZK distance of 50 mpc from us. It may be possible with current statistics to see a set of energy–angle correlations centered on the supergalactic plane (SGP).

The TA collaboration searched for such a set of energy-angle correlations.[3] On a 2° grid on the sky, many trials were examined using the set of variables listed above, and the correlation was scored using Kendall's correlation formula. The highest correlation was assigned to that point on the sky. The result is shown in Figure 9.6.

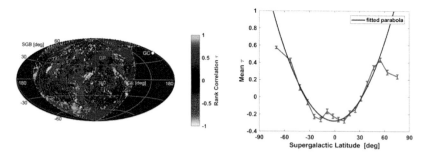

Figure 9.6 Energy-angle correlations in the TA data, in supergalactic coordinates. Left-hand panel: the supergalactic plane is labeled SGP, and negative energy–angle correlations, shown in red, cluster in its vicinity. Right-hand panel: the mean correlations as a function of supergalactic latitude, fit to a parabolic function. From J.P. Lundquist and P. Sokolsky, *Proceedings of Science PoS (ICRC2019)*, 2019, p. 343, with permission.

The left-hand panel of Figure 9.6 shows that negative correlations cluster in the region of the supergalactic plane. This means that events with higher energy are closer to the supergalactic plane than those of lower energy. The blue regions at higher supergalactic latitudes seem to show the same correlations; that is, the high-energy events are closer to the supergalactic plane. The right-hand panel shows the mean correlation parameter as a function of supergalactic latitude, fit to the function $ax^2 + bx + c$. The a-parameter is taken as indicative of the effect being searched for; that is, angle–energy correlations arising from magnetic fields in the body of the Virgo supercluster. One method for comparison with random Monte Carlo trials is to compare the a-parameter of each trial to the data. Using this method, the strength of the correlation is about 4.1 standard deviations.

References

[1] A. Aab et al., arXiv:1709.07321.
[2] R.U. Abbasi et al., *Ap. J.*, **790**, 2014, p. L21.
[3] J.P. Lundquist and P. Sokolsky, *Proceedings of Science PoS (ICRC2019)*, 2019, p. 343.

10

COMPOSITION
Direct Methods

10.1 Introduction

Direct methods of measuring the cosmic ray spectrum and composition are available for energies below 0.1 PeV. In this case, the primary cosmic ray particle interacts directly in the detector, and its nature and energy can be established using standard high-energy physics experimental techniques. These techniques are much more precise and reliable than the indirect methods necessary at higher energies. Correspondingly, precise data exist on spectra and composition up to about 100 TeV primary energy; recently, however, several experiments have extended this direct measurement domain up to 1 PeV.

Because of attenuation in the atmosphere, such measurements must be made at very high altitudes. While, historically, experiments were mounted at high mountain elevations, the most precise information has been garnered by flying detectors up to very small atmospheric depths (on the order of 1–10 g/cm^2) in balloons. The area and payload of such detectors is clearly limited. Recent technical improvements in controlling high-altitude balloons have enabled such balloons to stay aloft for much longer. In particular, the new class of super-pressure balloons can have flight times of many weeks, or even longer. Hence, detectors of the order of 1 m square in area can be exposed to cosmic rays for times on the order of a tenth of a year. Such balloon-borne experiments are attractive due to their relatively low cost in relation to space-based experiments. Small detectors have also been placed on satellites. More recently, two large detectors have been placed on the International Space Station (ISS) and exposure times of up to seven years have now been achieved.

10.2 The "Low Energy Composition"

From our point of view, the information on spectra and composition that such experiments can give provides the standard against which interesting deviations—due, for example, to the presence of an extragalactic flux at higher energy—can be measured. As direct data becomes available near the "knee" of the spectrum and EAS experiments on the surface extend their sensitivity to the "knee" from above, overlap of direct and indirect methods may be possible in the near future. This would allow very important cross-calibration of composition measurement, for example, as well as establishing a link between the direct and indirect energy scales. The spectrum and composition of 10–100 GeV/nucleon cosmic rays, for instance, can be called the "low energy composition," and can be considered to be representative of the galactic cosmic ray source abundances once allowances are made for particle fragmentation due to propagation effects. These abundances give significant clues to the origin and acceleration mechanisms of cosmic rays.

An important distinction is the one between "primary" and "secondary" nuclei. Primary nuclei (p, He, C, O) are produced in the source; secondary nuclei (Li, Be, B) can only be produced by secondary reactions of the primary nuclei. The ratio of "secondary" to "primary" fluxes as a function of energy is of great interest, as it gives us important information about the propagation and lifetime of cosmic rays in our galaxy.

While pioneering balloon experiments found that the low energy cosmic ray spectrum had the expected power law energy dependence even at low energies, direct measurements showed that interesting effects can occur. Figure 10.1 shows that, at energies of a few hundred GeV per nucleon, the Fe spectrum from early experiments was found to be somewhat flatter than that for protons and lighter nuclei. If one extrapolated to total energies of a PeV (around the "knee" of the total cosmic ray spectrum), then the iron flux would substantially exceed the proton flux.[1] One could conclude that galactic cosmic ray accelerators inject more iron into the interstellar medium as energy increases. Such a simple extrapolation is dangerous, however, because the relative abundance of secondary (spallation) products in the cosmic ray spectrum (such as lithium, beryllium, boron, etc.) is known to

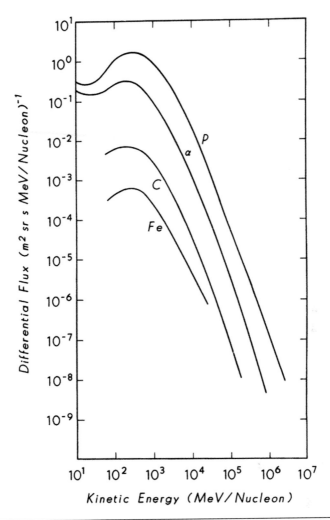

Figure 10.1 Fits to charge spectra for nuclei from p to Fe up to ~ 1 TeV/nucleon from early 1980s balloon experiments. The flatter slope of the Fe spectrum was considered controversial.

decrease as the total energy approaches the region of the "knee."[2] In other words, propagation effects are energy dependent. In general terms, the energy spectrum measured near the Earth may be different from the injection spectrum at the galactic accelerators because of energy losses during propagation, spallation reactions, and the escape of particles from the galaxy. However, energy losses are insignificant for particles heavier than 10 a.m.u.,[3] and the spallation cross section is known to be nearly energy independent.[4] The most likely reason for the spectrum difference is the charge-dependent increasing leakage of primary cosmic rays out of the galaxy with higher energy (as expected from the leaky box model).[5] The decrease of the secondary/primary ratio implies that cosmic rays must have an energy dependent total path length that decreases as one approaches the "knee." In other words, cosmic rays seem to traverse less interstellar matter as their energy increases.

10.3 Steepening of the Primary Spectrum

It is instructive to begin with a very simple picture, such as the one proposed by Muller et al.[6] Suppose the injection spectrum at the acceleration source has a power law of the form $I(E) = CE^{-\gamma}$ and $\rho(E)$ is the local cosmic ray density. If $\Lambda(E)$ is the cosmic ray path length before escape, and $\beta_i(E)$ is the mean free path for collisions for a given element; then for an equilibrium situation, we obtain Equation (10.1):

$$CE^{-\gamma_i} \propto \rho_i(E)\left[1/\Lambda(E)+1/\beta_i(E)\right] \tag{10.1}$$

where $\beta_i(E)$ has a slow energy dependence. $\Lambda(E)$ has an energy dependence that can be derived from the relative abundance of "secondary" spallation products as a function of energy, and which fits a power law between 1 and 100 GeV/a.m.u., as shown in Equation (10.2)

$$\Lambda(E) \propto E^{-0.5} \tag{10.2}$$

One can define an energy E_c for which $\Lambda(E) = \beta(E)$. At energies much greater than E_c, nuclear reactions become insignificant compared to propagation losses. The spectral shape then achieves the asymptotic form $\rho(E) \propto E^{-(\gamma + 0.5)}$.

Now $\beta(E)$ varies element by element, from 8 g/cm² for carbon to 2.5 g/cm² for iron (at these comparatively low energies). This means that E_c varies from a few GeV to 10–20 GeV/a.m.u. If the injection spectrum is the same for all nuclei, we expect all charge spectra to approach the same slope for $E >> E_c$. The point at which the turnover from low energy behavior occurs depends on E_c and is thus different for different nuclear species. As the Larmor radius of particles increases beyond a certain scale, escape from the galaxy may become even more rapid, and an additional steepening may result. However, the steepening at the "knee" may also be the result of approaching the maximum energy of the galactic accelerator. Such effects will modify the simple linear extrapolation and may not lead to a large fraction of Fe at the "knee." To understand this further, first, we examine a pioneering balloon-borne experiment and, then, move on to two contemporary experiments (balloon-borne and on the ISS).

10.4 The JACEE Experiment

The JACEE collaboration[7] was composed of a number of Japanese, U.S., and Polish institutions. Their effort had the goal of studying the charge spectra of cosmic rays up to 1 TeV/nucleon, and also to examine the details of nucleus–nucleus interactions at energies above those then available from accelerators. The experiment was a balloon-flown package of active and passive detectors. Typical flights reached altitudes of 3–5 g/cm^2. The basic idea was to measure the charge of the primary via a measurement of its specific ionization where $\beta = v/c$. The total energy was measured through calorimetry. Both measurements were taken using passive detectors such as emulsions, X-ray films, and CR-39 and Lexan etchable plastics. Some balloon flights also used Cherenkov counters and other active detectors in conjunction with the emulsion stack. The passive detector (see Figure 10.2) was divided into three basic sections: the charge detector, the target, and the calorimeter. The total detector was composed of 350 layers of alternating materials.

The charge detector was composed of emulsions 200–400 μm thick, as well as some plastic detectors. The specific ionization was measured using the time-tested emulsion techniques of grain density, grain gap distribution counting, and delta ray distributions.[8] The plastic detectors (CR-39 and Lexan) respond to the passage of heavily ionizing radiation by exhibiting a track of intensely damaged material along the particle trajectory. This track (about 50 Å in radius) has many broken long molecular chains and therefore is highly reactive when exposed to acid etching.[9] The resultant "etch cones" can be seen under a microscope, and can give information on the specific ionizations. This is particularly useful for $Z > 6$ nuclei. The error on charge determination using these techniques is 0.2 q_e for proton and He nuclei, and increases to about 2 q_e for Fe.

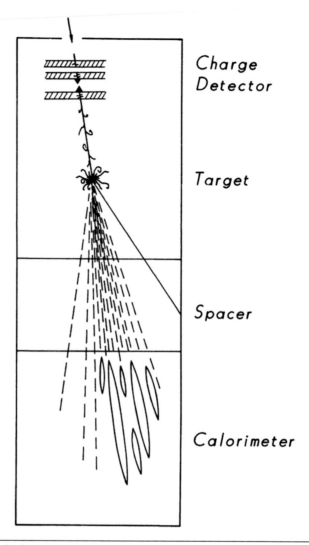

Figure 10.2   Schematic illustration of JACEE detector.

The target section was composed of thin emulsion plates (50–75 μm) alternating with acrylic sheets. This section represents a substantial mass, and most particles will interact here. The emulsions are used to determine the interaction vertex by following the charged tracks backwards, while the plastic plates are used to identify nuclear fragments from the interaction.

The calorimeter was seven radiation lengths deep and composed of thin lead sheets, alternating with thin emulsion plates and X-ray film that is sensitive to high energy γ-rays. Analysis of the X-ray film density allows a measurement of individual γ-ray energies. The energy resolution of the device can be measured by looking at the invariant mass distribution of pairs of γ-rays and finding the π_0 peak and its width. For individual γ-rays, the energy resolution is found to be 22%, and is somewhat larger for the sum of total observed γ-ray energy in an event.

The JACEE collaboration flew 14 flights from 1979 to 1996 and accumulated significant data. Their results, spanning an energy range from 2 TeV to 1 PeV, indicate no significant evidence for heavy nucleus dominance up to 1 PeV,[10] with abundances as follows: p:He:C-O:Ne-S:Z>18 16:29:35:9:10% with errors of +/− 5%. The proton and He flux was found to be consistent with a single power law, and the heavier nuclei exhibited power laws with harder spectral indexes.

A limiting problem for balloon-borne experiments that attempt to measure spectra above 100 GeV/nucleon is that the interstellar material traversed by the nuclei on their way to the Earth is on order of 1 g/cm², while measurements are taken at 3–10 g/cm² in the atmosphere. Such elevations mean that significant corrections must be made for atmospheric interaction before detection. At 7 g/cm², for example, more than half of the observed flux is generated in the atmosphere.[11] For this and other reasons, experiments at the ISS or on satellites have fewer such systematic corrections, as well as much longer lifetimes.

10.5 The Alpha-Magnetic Spectrometer

The Alpha-Magnetic Spectrometer experiment was the first cosmic ray detector on the ISS.[12] It is a powerful High-Energy-Physics type of magnetic spectrometer that is sensitive up to ~1 TeV particle energies. Its primary physics goals are to measure the anti-particle flux and to take precise measurements of the nuclear elemental energy spectra. It utilizes a suite of seven different detectors and a 0.14 Tesla permanent magnet (see Figure 10.3). The different detectors allow the measurement of the incoming particle energy and charge in multiple ways: from the particle curvature in a magnetic field (9 layers of Si microstrip detectors), from energy deposition, from the 20 layer transition

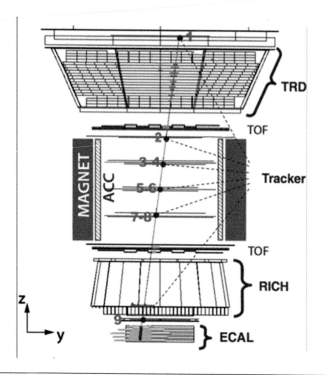

Figure 10.3 Schematic of the Alpha-Magnetic Spectrometer.[13]

radiation detector (TRD), and from the ring-imaging Cherenkov (RICH) ring radius detectors. The resultant charge resolution is ~0.1–.2 q_e. The discrimination between matter and anti-matter is provided by the spectrometer. This overall redundancy makes results from the Alpha-Magnetic Spectrometer very precise. The detectors were calibrated at CERN test beams at the appropriate energies. The instrument has a large aperture and has been taking data since 2011, giving its results not only the distinction of the highest statistical accuracy, but also good control over systematic error.

Some of the most important results from the Alpha-Magnetic Spectrometer have to do with the electron/positron energy spectra[14] (see Figure 10.4). The positron spectrum shows a surprising upturn at higher energies and clearly deviates from expectations of positrons produced in the interaction of ordinary cosmic rays as they travel through the galaxy. The high energy positron enhancement could be a signature of dark matter annihilation or unknown nearby

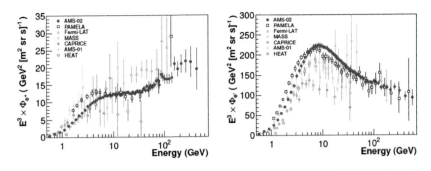

Figure 10.4 Results on the electron (right) and positron (left) fluxes. From C. Pizzolotto, *J. Phys.: Conf. Ser.*, 718, 042046, with permission.

astrophysical sources.[15] The enhancement is consistent with results from other lower statistics calorimetric experiments, but is much more precise.

The Alpha-Magnetic Spectrometer can measure the energy spectra of nuclei up to ~1 TeV, and has shown that the simple assumption of a uniform power law spectrum up to TeV energies does not hold (see Figure 10.5). Instead, the spectral power law index becomes the same at a rigidity (momentum/charge) near 60 GV for all the primary

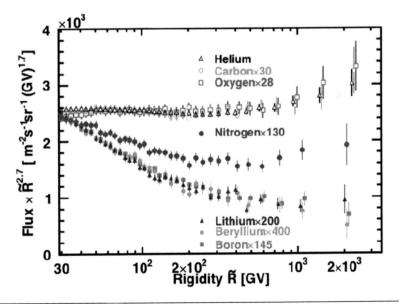

Figure 10.5 Alpha-Magnetic Spectrometer results for individual nuclear species as a function of particle rigidity $R = p/q$. From Pizzolotto, *Phys. Rev. Lett.*, 120, 2018, 021101, with permission.

nuclei (He, C, N, O), and then hardens above 100 GV. Secondary nuclei, produced by the spallation of primary nuclei during galactic propagation, also have departures from a uniform spectral index at roughly this rigidity, and their ratio with respect to a primary nucleus, such as C, decreases with energy. These precise results imply that our current understanding of galactic cosmic ray production and propagation is overly simplistic. Similar departures from simple power law behavior are observed at higher energies approaching the "knee" by the CREAM experiment and others (see Section 10.6).

10.6 The CREAM Experiment

The CREAM experiment consists of a program of balloon flights,[18] topped off by a pioneering ISS detector run, to measure the elemental composition of cosmic rays from 100 GeV to as near the knee at 1 PeV as possible. The experimental modules evolved over the course of the balloon flights, but they basically consisted of a ~2 m × 1 m × 1 m detector that measured the charge of the incoming particle, allowed it to interact in a carbon target, and used a calorimeter subsequently to measure the energy of the primary. The total exposure time of the balloon flights was 191 days, with typical single balloon flights launched in Antarctica lasting 20–40 days. The balloons reached altitudes corresponding to an atmospheric overburden of ~4 g/cm². The right-hand side of Figure 10.6 shows a schematic of a typical CREAM detector. Charge measurement is performed with timing charge detectors (TCDs), silicon charge detectors (SCDs) and transition radiation detectors (TRDs); the calorimeter is composed of interleaved scintillating fiber and tungsten plates. The detector has an aperture $A\Omega$ of ~2.2 m²sr. The detector achieves charge resolution of $0.2q_e$ and an energy resolution of ~40% as calibrated with charged particle beams at CERN. The power of good charge resolution can be seen on the left-hand side of Figure 10.4, which shows the charge distribution of galactic cosmic rays from proton to Fe and above.

In 2017, a CREAM module was launched on a Space-X rocket that docked with the ISS. The ISS–CREAM detector (see Figure 10.7) is similar to the balloon-borne payloads, but has some newer features. The charge measurement is performed with silicon charge detectors (SCDs), and a boron doped scintillator (BSD) at the bottom of the module

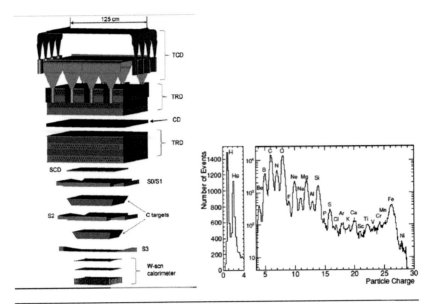

Figure 10.6 Typical CREAM detector balloon package (left-hand side) and measured cosmic ray charge distribution (right-hand side). From H.S. Ahn et al., *Ap. J. Lett.*, 714, 2010, L89, with permission.

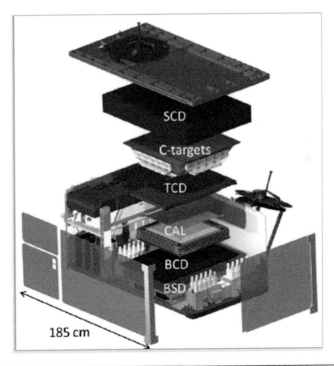

Figure 10.7 ISS–CREAM configuration. From E.S. Seo et al., *Adv. Space Res.*, 53, 2014, 1451, with permission.

captures thermal neutrons from the cascade in order to improve electron/proton separation.

10.7 Results from CREAM

Figure 10.8 shows the result from several flights of the CREAM detector.[21] The most important finding is the apparent difference in spectral index between protons and He and the heavier elements, as well as a hardening of the spectrum at an energy near 200 GeV/nucleon. The elemental spectra deviate from the simple power law spectrum seen at lower energies. This is quite surprising if the sources of galactic cosmic rays are super novae (SN) in the steady-state continuous distribution approximation, which should generate power law spectra with similar indexes for all the elements up to Fe. There is speculation that spectral structure can be induced by in-homogeneity of sources in time and space modulating Z-dependent escape from galactic confinement. The spectral structure may represent the importance of recent nearby super novae contribution to the local cosmic ray flux. One would expect more recent sources to dominate over older ones at the highest energies due to propagation energy losses.

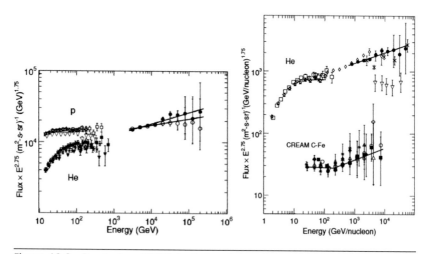

Figure 10.8 Results from CREAM: hardening of elemental spectra above 200 GeV/nucleon (right); the difference in spectral index for protons and He (left): the figure includes lower energy results from Alpha-Magnetic Spectrometer and other experiments. From H.S. Ahn et al., *Ap. J. Lett.*, 714, 2010, L89, with permission.

10.8 Conclusions

Recent precision measurements of primary and secondary cosmic ray nuclear spectra have upended the simple picture of super novae shock acceleration producing a single power law spectrum that is only modified by the effects of the leaky box model of escape from the galaxy. The differences between proton and He, C and O spectra and the hardening of the spectral indexes at energies above 100–200 GV will no doubt lead to a more refined and complex picture of galactic cosmic ray acceleration and propagation. Detailed studies of isotopic composition and heavy nuclear abundances beyond Fe are also important in addressing this picture, but are beyond the scope of this introductory volume.

References

[1] D. Muller, *Workshop on Ultrahigh Energy Cosmic Ray Interactions*, Philadelphia, PA, 1982, 448.

[2] J.F. Ormes and R.J. Protheroe, *Astrophys. J.*, **272**, 1983, 756–764.

[3] D. Muller et al., *Proceedings of the 20th ICRC*, Moscow, Russia, 1987, **vol. 1**, 334.

[4] R. Silberberg and C.H. Tsao, *Astrophys. J.*, **35**(Suppl. Ser), 1977, 129.

[5] J.F. Ormes and R.J. Protheroe, op. cit., p. 759.

[6] D. Muller et al., *Proceedings of the 20th ICRC*, Moscow, Russia, 1987, **vol. 1**, 334.

[7] T.H. Burnett et al., *Workshop on Very High Energy Cosmic Ray Interactions*, Philadelphia, PA, 1982, 221; T.H. Burnett et al., *Phys. Rev. Lett.*, **51**, 1983, 1010.

[8] C.F. Powell, D.H. Fowler, and D.H. Perkins, *The Study of Elementary Particles by the Photographic Method*, Pergamon Press, Oxford, 1959.

[9] M.S. Longair, *High Energy Astrophysics*, Cambridge University Press, Cambridge, 1981, 102.

[10] Y. Takahashi et al., *Nuclear Phys. B*, **60**(3), 1998, 83–724.

[11] M. Simon et al., *Astrophys. J.*, **239**, 1980, pp. 712–724.

[12] R. Battiston, *Nuclear Instr. Meth.*, **588**, 2008, 227.

[13] M. Aguilar et al., *Phys. Rev. Lett.*, **110**, 2013, 141102.

[14] M. Aguilar, *Phys. Rev. Lett.*, **113**, 2014, 121102.

[15] M. Turner and F. Wilczek, *Phys. Rev. D*, **42**, 1990, 1001.

[16] C. Pizzolotto, *J. Phys.: Conf. Ser.*, **718**, 042046.

[17] M. Aguilar et al., *Phys. Rev. Lett.*, **120**, 2018, 021101.

[18] E.S. Seo et al., *Astropart. Phys.*, **39–40**, 2012, 76-87.

[19] H.S. Ahn et al., *Ap. J. Lett.*, **714**, 2010, L89.

[20] E.S. Seo et al., *Adv. Space Res.*, **53**, 2014, 1451.

[21] H.S. Ahn et al., *Ap. J. Lett.*, **714**, (2010, L89.

11

COMPOSITION

Muon and Electron Detectors

11.1 Introduction

Because of the low flux, composition studies in the energy range above 1 PeV must be indirect at the present time, utilizing information derived from the EAS, rather than the primary particle itself. This chapter examines the extent to which the study of the muon and hadron components of the extensive air showers (EAS) can lead to a better understanding of the primary composition.

Muons produced in EAS can be used to study the chemical composition of the primary cosmic ray flux as their multiplicity depends on the atomic number of the primary particle. From a practical point of view, the study of muons in EAS breaks up into two modes: surface detection of predominantly low energy (~1 GeV) muons over a large area, and deep underground muon detection of high energy (≥ 100 GeV) muons over a small area around the shower core direction. We briefly review relevant facts about the muon content of EAS, and then describe the two different experimental techniques used to study composition.

11.2 Muons in the EAS

The probability that a pion or a kaon will decay depends on:

- its energy, as this determines the decay length $c\gamma\tau$, where c is the speed of light, γ is the relativistic dilation factor, and τ is the lifetime of the particle in its rest frame;
- the available decay length before the particle hits the Earth's surface;
- the density of the atmosphere as this determines the likelihood that the particle will interact before decaying.

Due to this competition between interaction and decay as the pion and kaon energy increases, fewer particles decay. For pion and kaon energies of less than 100 GeV, the decay probability is large, and the resultant single muon energy spectrum has the same form as the primary cosmic ray spectrum.[1] For higher energy particles, the muon spectrum steepens by one power of the energy. While it is true that muons are generated primarily from the decay of π mesons (pions) and K mesons (kaons), a small fraction of high energy muons are the result of direct (also known as "prompt") production processes such as D meson decays. Muons from prompt production have a harder energy spectrum than those from pion and kaon decays.

The highest energy muons in an EAS are the result of the decay of the highest energy pions and kaons and, therefore, carry information about the early development of the shower. These muons are found very near the shower axis, are produced between 8 and 16 km above observation level, and diverge from the shower axis in approximately straight lines.[2] Their transverse momentum distribution can be used to determine the transverse momentum distribution, $P_\perp$, of their parent pions and kaons. As will be seen, underground experiments measuring such muons have established that the average transverse momentum of hadrons increases from about 400 MeV/c at low energies to 500–600 MeV/c at TeV to PeV energies.[3] This trend in the primary interaction has been confirmed by collider experiments.[4] The number of these high energy muons in an EAS is more sensitive to primary composition than to the details of hadronic interactions. However, limitations on the collecting areas of deep underground detectors make this technique only useful at lower energies (around 1–10 PeV).

11.3 The Muon Lateral Distribution

The muon lateral distribution depends on the minimum muon energy to be considered. For typical ground arrays, this is ~1 GeV or lower. In general, the number of muons is 50–100 times smaller than the number of electrons near the shower axis, but these become the dominant component at distances of 1 km or greater. Although the muon lateral distribution is much flatter than the electron lateral distribution, the muon density remains low, while the average muon energy is much higher than the average electron energy. This has important implications for detector design.

In the past, detectors that study low energy muons from EAS (on the order of one GeV) did so using counters buried under at least 10 feet of earth (or equivalent shielding material). This effectively absorbs the hadronic, electron and γ-ray component of the EAS. Such detectors can study the muon lateral distribution and the muon multiplicity. These measurements are compared with a simultaneous measurement of the electron size and lateral distribution on the surface. As the muon density is low (typically $\rho_\mu/\rho_e = 0.1$ at 200 m from the EAS axis), plastic scintillators used for muon detection must be on the order of ten times larger in area than the electron counters in order to have similar statistical accuracy (due to Poisson fluctuations) in determining density. On the other hand, the flat muon lateral distribution means that counters can be widely dispersed without significantly affecting minimum primary energy triggering thresholds for the muon array.

More recently, improvements in the time resolution of scintillation and water Cherenkov detectors have made it possible to study the muon component at large distances from the shower axis via the characteristic fast single minimum ionizing signals produced.[5] Large water Cherenkov detectors are also more sensitive to muons than to electrons, since muons will typically traverse the entire detector while electrons will be absorbed.[6]

11.4 High Energy Muons

High energy (order of hundreds of GeV to TeV) muons are measured in deep underground detectors. Many proton decay experiments have been used to study such muons; results have been obtained from Sudan I,[7] IMB,[8] Freijus,[9] and NUSEX[10] and the like. More recently, the IceCube[11] experiment at the South Pole has added the study of such muons to its astrophysics program. The minimum detectable energy is determined by the overburden of matter. Different detectors are easily comparable if the overburden is converted into an equivalent depth of water, known as the "meters water equivalent" (mwe). Typical installations range from 1000 to 4200 mwe depths.

Since most such high energy muons are very near the shower axis, very large detection areas are not needed to measure muon multiplicities at PeV energies even though reasonably fine detector segmentation is required. If, for example, a detector is located at a depth of 4200 mwe, which corresponds to an minimum muon energy E_{min} of 2.7 TeV then, to measure the multiplicity in an EAS, the detector should be large enough to see all the muons produced with energies down to E_{min}. We can assume that pions and kaons will be produced with $P_\perp$s up to 1–1.5 GeV/c. The perpendicular detector length scale can be estimated from the expression given in Equation (11.1):

$$L = P_\perp \left(h/E_\pi \right) \qquad\qquad (11.1)$$

where h is the altitude of the primary interaction, and E_π is the parent pion energy. For $h = 19$ km and $E_\pi = 3.6$ TeV, $L = 2.6$ m. Therefore, a 10×10 m detector will easily cover a $P_\perp$ range out to 2 GeV/c.

11.5 Muon Multiplicity Depends on Composition

The number of low energy (order of GeV) and high energy (order of TeV) muons in an EAS depends on the atomic number of the primary particle. This can be seen most simply by comparing what one expects for protons and iron primaries. In the simple superposition model of nuclear interactions, an Fe nucleus breaks up and produces 56 individual EAS, each of which is generated by a nucleon of energy $E_0/56$. The resulting pions in each EAS will have smaller average energies than they would for a proton induced EAS of the same energy. More of such low energy pions will decay to muons before interacting because the decay length has decreased, while the interaction length is approximately the same. One may also expect than the pion multiplicity will be larger for Fe initiated showers. One can see this from the following simple argument.[12] If the pion multiplicity for nucleon interactions increases as $n(E) \propto E^a$, where $0 < a < 1/2$, then for the individual nucleons in Fe interactions $n(E) \propto (E/A)^a$ and, hence, the total multiplicity will follow the form $n_{Fe}(E) \propto A^b(E/A)^a$, where $2/3 < b < 1$, or, in the event that $a = 0.5$ and $b = 1$, the ratio of Fe to p multiplicity will follow the form $n_{Fe}/n_p \propto A^{.5}$.

Even with a slow multiplicity increase with energy, heavy nuclei are expected to generate more pions. For reasonable hadronic interaction models, Fe interactions at PeV energies are expected to yield on the order of two times more low energy muons at the surface than would be the case for protons. A rough estimate of the number of high energy muons in an EAS produced by a particle of atomic number A and energy E_0 is given by ,[13] Equation (11.2):

$$N_\mu\left(>E_{min}\right)=KA\left(\sec\vartheta/E_{min}\right)\left(E_{min}/\left(E_0/A\right)\right)^a\left(1-E_{min}/\left(E_0/A\right)\right)^b \quad (11.2)$$

E_{min} is in GeV, $\alpha = -0.757$ and $b = +5.25$, while $K = 14.5$.

This result is based on Monte Carlo calculations assuming scaling and holds for zenith angles of less than 60 degrees.

11.6 Composition Determined from Muon Multiplicity

The experimental method for determining composition using muon multiplicity is, in principle, simple. One measures the number of muons N_μ and compares with Monte Carlo simulated expectations for a given cosmic ray spectrum, composition, and hadronic interaction model. However, it is not straightforward to interpret this as, for a given minimum muon energy E_{min}, the primary energy for an Fe nucleus to produce such muons is higher than the primary proton energy needed for the same E_{min}. In other words, since energy per nucleon is the relevant variable for muons, once E_{min} is set, much higher total energy Fe nuclei are required to achieve this E_{min} than for protons. The net result is that one does not compare proton and Fe fluxes at the same energy. For example, for E_{min} on the order of one TeV, proton primaries will, on average, produce 1 muon, while iron nuclei will generate more than 1. Because of the different threshold energies for generating 1 TeV muon, comparing $N_\mu > 1$ muon to $N_\mu = 1$ muon rates relates the integral flux of protons with energy < 1 PeV to the integral flux of Fe above 1 PeV.

Since the low energy flux is much larger, small uncertainties can easily be magnified through spill down effects. For this reason, a much more sensitive way to determine composition is to measure N_μ for fixed primary energy E_0 (assuming E_0 is large enough so that both protons and Fe nuclei have similar detector efficiencies). In practice, one combines data into energy bins whose width is set by the detector energy resolution. For surface detectors, this devolves into measuring N_μ for fixed intervals of energy derived from electron size N_e measured at the surface. Of course, N_e and E_0 have large fluctuations relative to each other. Furthermore, since Fe nuclei interact higher in the atmosphere than protons, Fe and p nuclei of the same energy will not produce the same size N_e at the surface. All these effects must be carefully simulated, including the resolutions imposed by the nature of the detectors used, before conclusions regarding composition can be made. The energy dependence of various processes in the simulation now become particularly important. For higher energies, better results are found by using S800, or an equivalent parameter, to establish the shower energy, as this has a lesser dependence on the composition of the primary particle. A major advantage for hybrid detectors is that the air fluorescence detector measures the energy completely independently of the surface array. As we shall see, for all types of experiments, current hadronic models and simulations of EAS have difficulty in reproducing experimental muon multiplicity results.

In the following sections, we describe the methodology and results from the IceCube/IceTop, HiRes/MIA, and the Auger experiments.

11.7 The IceCube/IceTop Measurement of Composition

The IceCube/IceTop experiment[14] is unique in that it can measure three components of an EAS: the electron size at the surface using a 1 km² water Cherenkov array; the muon multiplicity far from the core at the surface, using the same array; and the TeV muon multiplicity in the deep (1.5–2.5 km depth in ice) km3 IceCube detector (see Figure 11.1). Photomultipliers in the deep ice detect Cherenkov radiation from high energy muons as well as showers of particles from neutrino interactions in or near the active volume of the detector. The multiplicity of muons in the bundle around the shower core reaching the deep detector is estimated from the energy deposition in the ice. Coincidence measurements of TeV muon multiplicity with surface muon density and energy derived from the electron size at the surface also allow a check on the adequacy of the hadronic model used to simulate the interactions in Monte Carlo. This experiment has sufficient aperture to cover the cosmic ray energy range from 1 PeV to ~1 EeV.

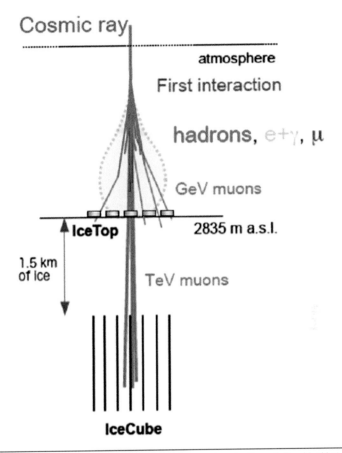

Figure 11.1 Schematic layout of IceCube/IceTop experiments. From H. Dembinski et al., *EPJ Web of Conf.* 145, 2017, p. 01003, with permission.

Current results from this experiment indicate a puzzle. Combining surface detector information with TeV muon multiplicity and using a neural net approach yields composition results at 100 PeV–1 EeV that are in conflict with air fluorescence measurements. Both Auger and TA find a light composition near 1 EeV, while TeV muon multiplicity indicates a dominantly Fe composition. Furthermore, using GeV surface muon multiplicities *alone* similarly indicates an increasingly heavy composition that even a pure Fe flux cannot explain. This indicates a problem in our understanding of how to model the muon content of EAS, both for low energy and high energy muons. Similar problematic results were found by the much earlier HiRes–MIA experiment at similar energies, but using air fluorescence to determine the shower energy.

11.8 Hybrid Air Fluorescence Arrays

11.8.1 HiRes/MIA

The most reliable measurements of low energy muons as a measure of composition are performed by hybrid air fluorescence and surface detector arrays. The HiRes/MIA experiment[16] (see Figures 11.2 and 11.3) pioneered this approach and preceded all current hybrid experiments. It utilized a prototype HiRes air fluorescence detector that overlooked a 1 km² plastic scintillator array (CASA) located 3 km from the fluorescence detectors. Buried under the array at a depth of 10 m was another array of scintillators (MIA) that detected > 1 GeV muons. While this CASA/MIA array was originally designed to search for gamma ray flux from point sources, in which

Figure 11.2   HiRes/MIA experiment. Surface detector (left) and HiRes prototype (right). The array is centered on the original Fly's Eye II detector. From P. Sokolsky, with permission.

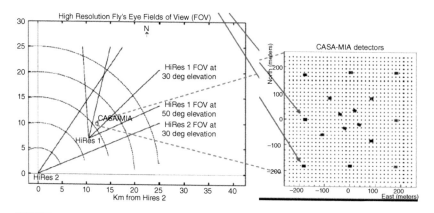

Figure 11.3   HiRes/MIA layout. From T. Abu-Zayyad et al., *Nucl. Inst. Meth. A*, 450, 2000, p. 253, with permission.

case the muon counters would serve as a veto for high multiplicity hadronic EAS, the veto counters could be used to estimate the number of muons, while the HiRes detector provided a reliable measurement of the EAS energy. The detector operated in a unique fashion in this mode. The core of the EAS was not constrained to lie in the CASA/MIA array since the core position could be determined by the HiRes event plane and the timing information from the MIA array. The MIA counters then simply sampled the muon lateral distribution at different distances for different events. In this way, the effective aperture of the detector was increased from the nominal 1 km^2 to ~10 km^2. The experiment ran out of statistics near 1 EeV. Results indicated an excess of muons even for an assumed pure Fe flux.

11.8.2 Pierre Auger Observatory

The Auger surface array[18] is very large, at ~3000 km^2, and in this case the EAS core is constrained to lie within the surface array boundary. Muons dominate the signal in the Auger water Cherenkov detectors in general and, at distances of ~1 km from the shower core, they dominate the particle flux hitting the detectors, too. This, combined with timing information (which can, in many cases, identify muons by their distinctive sharp time profile) allows this experiment to have significant sensitivity to the muon component of EAS.

The first hint of a problem with muons in this experiment came from comparing the surface detctor signals for hybrid events to predictions from EAS simulations. The experimenters chose a fixed distance (1000 m) from the core of each shower to which to interpolate the lateral distribution. The distance is chosen to minimize fluctuations from shower development and composition dependence. The relation between this signal value and the simulated signal at the same primary energy is composition dependent. The Auger experiment finds that no hadronic model can give good agreement between the measured signal size and the simulated size, not even for the case of a pure Fe flux. The overall disagreement is a ~factor of 1.3. Since muons dominate the Auger signal overall, it is believed that this implies an excess of muons in real events compared to simulations, even if one assumes extremely heavy compositions.

11.8.3 The Telescope Array Detector

The TA surface detector,[19] composed of thin plastic scintillators, has equal sensitivity to electrons and muons; thus, it is only sensitive to the muon content at large distances from the shower core. This, nevertheless, can be exploited to study muon density. Because of the dominance of the muon lateral distribution at these distances and the absorption of the electron component by the atmosphere for highly inclined events, selecting large zenith angle events and counters that are far from the shower core enhances the sensitivity to muons significantly. At distances more than 2 km from shower core and with such cuts, simulations show that the muon signal dominates the electron signal. Measuring the shape of the lateral distribution sampled by such counters thus has good sensitivity to the muon component. But, once again, the TA analysis shows that no hadronic model can match the observed number of muons, even for the case of a pure Fe composition.[5] The overabundance of muons compared to simulations is consistent with the Auger result.

11.9 Implications of the Muon Excess

Unless there is unknown but significant systematic error in all the experiments described above, it appears that a muon excess exists over a wide range of primary energies, from PeV to EeV. The muon excess also seems to be observed both with low energy muons (HiRes/MIA, Auger, and TA) and with TeV scale muons (IceCube/IceTop), though the latter is only determined at PeV energies. This would seem to rule out an energy threshold effect where unknown new physics could enrich muon production. In any case, low energy muons at large distances from the shower core are produced by mundane relatively low-energy pion and kaon decays. It may still be the case that the implementation of better knowledge of lower energy production cross sections will resolve this puzzle.

References

[1] T.K. Gaisser and G.B. Yodh, *Ann. Rev. Nucl. Part. Set.*, **30**, 1980, p. 491.
[2] K. Greisen, *Ann. Rev. Nucl. Sci.*, **10**, 1960, p. 81.

[3] M.L. Cherry et al., in *Workshop on Very High Energy Cosmic Ray Interaction*, Philadelphia, 1982, pp. 278–289; J.W. Elbert, in *Workshop on Very High Energy Cosmic Ray Interaction*, Philadelphia, 1982, p. 312; T.K. Gaisser and T. Stanev, in *Workshop on Very High Energy Cosmic Ray Interaction*, Philadelphia, 1982, p. 295.

[4] J. Whitmore, *Phys. Rep.*, *10C*, 1974, p. 273; K. Apgard et al., *Phys. Letters*, *121B*, 1983, p. 209; G. Arnison et al., *Phys. Lett.*, *123B*, 1983, p. 108.

[5] R.U. Abbassi et al., *Phys. Rev. D*, **98**, 2018, 022002.

[6] A. Aab et al., *Phys. Rev. D*, **91**, 2015, 032003.

[7] J. Bartelt et al., *Phys. Rev.*, **D32**, 1985, p. 1630.

[8] R. Bionta et al., *Phys. Rev. Lett.*, **51**, 1983, p. 27.

[9] Ch. Berger et al., *Proc. 19th ICRC*, La Jolla, USA, 1985, vol. **8**, p. 257.

[10] G. Battistoni et al., *Phys. Lett.*, **133B**, 1983, p. 454.

[11] M.G. Aartsen et al., *Phys. Rev. D* **100**, 2019 082009.

[12] J. Stone et al., *Proc. 19th ICRC*, La Jolla, U.S.A, 1985, vol. **8**, p 226.

[13] D.S. Ayres, in *Third Workshop on Grand Unification*, Chapel Hill, 1982, p. 289.

[14] J.G. Gonzalez et al., *J. Phys.: Conf. Ser.*, **718**, 2016, 052.

[15] H. Dembinski et al., *EPJ Web of Conf.* **145**, 2017, p. 01003.

[16] T. Abu-Zayyad et al., *Phys. Rev. Lett.* **84**, 2000, 4276; T. Abu-Zayyad et al., *Ap. J.* **557**, 2001, 686.

[17] T. Abu-Zayyad et al., *Nucl. Inst. Meth. A*, **450**, 2000, p. 253.

[18] A. Aab et al., *Phys. Rev. D*, **93**, 2015, 019903; A. Aab et al., *Phys. Rev. Lett*, **117**, 2016, 192001.

[19] T. Abu-Zayyad et al., *Nucl. Inst. Meth. A*, **689**, 2012, p. 87.

ULTRAHIGH ENERGY COSMIC RAY COMPOSITION

12.1 Introduction

This chapter concerns itself primarily with the study of the > 0.1 EeV cosmic ray composition using the air-fluorescence technique. It is clear that the nature of that composition is pivotal in determining the origins of cosmic rays, not just because it gives clues to the chemistry of the source. As an example of the impact a measurement of composition can have, let us suppose that a heavy composition, dominated by Fe, is found at > 0.1 EeV. There are then a number of consequences:

1. If the composition is heavy, then the power law spectrum up to 10 EeV energies has many more candidate source models. Direct acceleration mechanisms may very well be able to account for such a flux *even* if coming from galactic sources, while many more extragalactic acceleration models can account for the > 10 EeV flux.
2. A heavy spectrum would explain the near complete isotropy observed up to 10 EeV and the weak anisotropy above that energy.
3. A cutoff at 60 EeV would mostly be due to photospallation on the cosmic microwave background if the sources were sufficiently distant.
4. The neutrino flux associated with the GZK effect would be very significantly reduced.

Unfortunately, this crucial issue can only be studied by indirect techniques. All attempts at measuring the composition that will be discussed in this chapter are based on the fact that $\sigma_{p-air} < \sigma_{Fe-air}$. It follows that the interaction length of iron (λ_{Fe}) is shorter than that for protons (λ_p). Iron will thus, on average, interact higher in the atmosphere. The air shower produced by an iron primary can be thought of as a superposition of 56 proton-like showers, each with ~1/56 of the primary energy. Thus, fluctuations in shower development will also be smaller for iron than for protons. Alpha particles—CNO, silicon, and so on—presumably also exist in the cosmic ray flux and will fall in between these two extremes.

12.2 The X_{max} Technique

In principle, one would like to measure the distribution of X_1, the point of first interaction of the cosmic ray, and compare this distribution with expectations for light and heavy composition. This is impossible with the air-fluorescence technique as so few particles are produced at this point and, thus, there is too little light. Instead, one measures the distribution of X_{max}, the depth of shower maximum, which is the brightest part of the cascade. Note that the distribution of X_{max} of an ensemble of events is the convolution of the distribution of X_1 with shower development fluctuations. The distribution of X_{max} relative to X_1 is model dependent; that is, if the interaction multiplicity increases rapidly with energy, $X_{max} - X_1$ will get smaller with energy. There is also reliance on expectations about the total cross section and inelasticity energy dependence. In the end, inferring composition from measured X_{max} distributions depends, to some extent, on assumptions about hadronic physics in the fragmentation region.

We note that the center of mass energy available at the large hadron collider at CERN (LHC) corresponds to a cosmic ray energy of 0.1 EeV. It is thus, in principle, possible to validate hadronic models used in composition analysis at this energy. However, most experiments at the LHC do not have acceptance in the forward fragmentation region, so some extrapolation is still necessary. Similarly, data on nucleus–nucleus collisions now exists from the relativistic heavy ion collider (RHIC) as well as the (LHC) and is aiding in calibrating the models. Nevertheless, significant extrapolations must be made to simulate data near 100 EeV.

The general problem can be stated in the following way. A variety of experimental data show that X_{max} at about 1–10 EeV is on the order of ~750 g/cm² with a fluctuation of 50–60 g/cm². Protons, with rising $\log^2 S$ cross sections and mild scaling violation (QGSJet 04 model), yield X_{max} of 780 g/cm^2[1], while He, N, and Fe nuclei would have mean X_{max} values of 760, 720, and 690, respectively. However, the uncertainties in hadronic models from one extreme to another are on the order of 30–40 g/cm^2 for protons at the highest energies. While it is already clear that a pure Fe composition is ruled out at this energy, a mixture of light and medium nuclei could still be consistent, as would a pure proton composition, taking into account systematic errors in the data and uncertainties in the models. For example, hadronic models such as EPOS can increase X_{max} by as much as 30 g/cm^2. Interpreted with this model, this data could be compatible with a larger medium and heavy nucleus component. The effect of hadronic model choice on X_{max} fluctuations is much smaller. In general, the shape of the X_{max} distribution is less sensitive to hadronic model assumptions and is therefore a more reliable measure of composition than the absolute value of the mean X_{max}. While some mixture of nuclei could mimic a proton distribution for the bulk of the data, the tail of the distribution (if it exists) is impossible to simulate using heavier nuclei. Thus, the width and the asymmetry of the X_{max} distribution is the most reliable indicator of composition.

We note here that the elongation rate of X_{max} tells us something about whether or not the cosmic ray composition is changing. X_{max} is a linear function of $ln\ E$ for a given atomic number A. For a mixed composition, one can apply the superposition principle and find that X_{max} is a linear function of $ln\ E$. One can thus look for changes in composition as a function of energy by studying the constancy of the elongation rate. It can be shown that, for a mixed composition, the expression for the elongation rate must be modified as shown in Equation (12.1):[2]

$$D_e = (1-B)X_0\left(1-\partial ln\langle A\rangle/\partial lnE\right) \tag{12.1}$$

where $\langle A\rangle$ is the average mass number. This expression is more sensitive to a change in the composition than to the details of the hadronic interaction model. For example, if the composition changes from pure Fe to pure protons in one decade, the elongation rate would become 184 g/cm^2, or more than double the rate observed at high energies. A change in hadronic physics would produce a much smaller effect, as most parameters are expected to change smoothly and slowly with energy.

12.3 Results from Air Fluorescence

The Fly's Eye type of detectors measures the shape of the longitudinal development on a shower by shower basis, and can determine both X_{max} and the rate of rise and fall of the shower.[3] Viewing showers in stereo from two eyes is an important cross-check for this application, as it allows a constraint on how well the reconstruction of the shower shape is done. Monte Carlo acceptance calculations for this technique show that the acceptance in X_{max} is not uniform. For a typical detector covering an elevation angle of 3–45°, it rises linearly below 500 g/cm², and then becomes relatively flat up to 1000 g/cm². The flat region neatly spans the expected region of X_{max} for $E > 0.1$ EeV. The resolution in X_{max} is ~15–20 g/cm² for events reconstructed in stereo or in hybrid mode, and is approximately independent of X_{max} in this range.

As finite resolution and acceptance effects can shift and distort the X_{max} distributions, they must either be corrected for, or minimized by appropriate data selection. In the first approach, one compares the experimental distribution to predicted theoretical distributions that have been modified to reflect these effects through detector simulations. This is done by generating EAS for a particular composition and hadronic model, passing them through the Monte Carlo detector simulation program, and producing simulated events that can be reconstructed using the standard analysis programs. The resultant X_{max} distribution incorporates the biases and resolution of the detector to the extent that they are understood and properly modeled. One way to check that this is correctly handled is to compare the real distributions with the Monte Carlo distributions in all available geometrical parameters (Rp, zenith angle, azimuthal angle etc.). For stereo and hybrid data, the agreement is typically very good. Figure 12.1 shows the X_{max} distribution for pure Fe, CNO, He, and pure proton fluxes for an energy interval near 3 EeV, with resolution folded in for a particular hadronic model (QGSJet II-04). There is a clear difference in the average X_{max}, width of X_{max} distribution, and presence or absence of an exponential tail at large X_{max} for these cases. The data shows that, at this energy, agreement with a pure proton flux is good.

For stereo data, the X_{max} distribution can be used to determine the X_{max} resolution experimentally, since the X_{max} difference distribution is

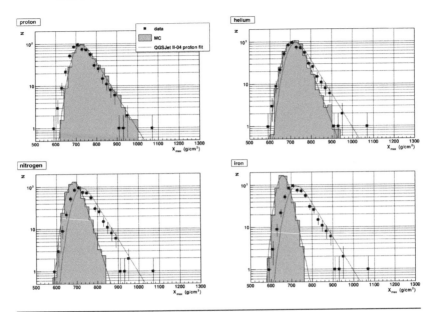

Figure 12.1   Predicted X_{max} distributions for p, He, CNO and Fe fluxes at an energy of ~ 3 EeV based on the QGSJet04 hadronic model. Also shown are TA data normalized to the same area. From R.U. Abbasi et al., *Ap. J.* 858, 2018, 76, with permission.

the convolution of the two resolution functions. It can also be directly compared to the simulated event difference distribution to check the simulation accuracy.

Figure 12.2 shows the predicted elongation rate for different composition assumptions as well as the result from the TA experiment in the range 1–10 EeV. Historically, the measurement of the elongation rate of X_{max} and the average fluctuation in X_{max} have been important to inferring composition. However, the shape of the X_{max} distribution—in particular, its rapid rise and long tail—are additional new pieces of information. For instance, a two-lobed distribution can have the same mean and standard deviation as a gaussian distribution, yet it has a very different implication for composition. The rise of the distribution is sensitive to the heavy component of the composition, while the long exponential tail reflects the presence of protons and alpha particles. The measurement of the falling slope of the distribution can be used to estimate the $\sigma_{p\text{-}air}$ inelastic cross section. One can demand that the position of the rise, the slope, and the magnitude of the tail, as well as $\sigma(X_{max})$, all be consistent with a particular cosmic ray composition.

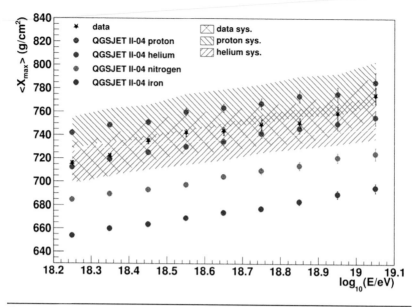

Figure 12.2 Elongation rate for different nuclei in the QGSJet-04 hadronic model. Data is from the TA hybrid experiment. From R.U. Abbasi et al., *Ap. J.* 858, 2018, 76, with permission.

From this point of view, all air fluorescence experiments agree that the data requires a significant proton contribution to the cosmic ray flux up to 10 EeV energies. Above this energy, the situation is still unclear. The Auger hybrid data shows a narrowing in the X_{max} distribution, suggestive of a heavier composition, while the HiRes stereo data was still consistent with a light protonic distribution. The TA hybrid data does not yet have the statistical strength to differentiate clearly in this energy range.

12.4 Composition Studies Using Cherenkov Light

Cosmic ray studies at lower energies (in the 10^{15}–10^{17} eV decades) are also possible using the X_{max} technique. Another way of describing the accuracy of X_{max} measurements is to quote the uncertainty in psi. If the uncertainty in psi is large (~5°, as for monocular fluorescence recon-struction), the geometrical reconstruction of events is not sufficiently accurate to yield unbiased measurements of X_{max}. If the uncertainty is small (1° or less, as for stereo or hybrid reconstruction), the geometri-cal reconstruction is sufficiently good enough to identify biases and improve reconstruction techniques to eliminate them.

The Telescope Array Low Energy Extension (TALE) has been built with fluorescence detectors that look at higher elevation angles (from 30°–60°), and a new class of events was found in TALE data: numerous events dominated by Cherenkov radiation. Cherenkov radiation is generated in a narrow cone (several degrees wide) around the EAS core and is very bright. Excellent reconstruction is possible for these Cherenkov dominated events, and the psi value cannot stray far from the correct value. Uncertainties in the ~0.5° range are typical. This is an unusual situation where small X_{max} biases are achieved in monocular mode. The detectors at the high elevation Auger telescopes (HEAT) are similar to those at the TALE and use a similar approach.

Figure 12.3 shows preliminary TALE $<X_{max}>$ results as shown at the ICRC 2019 conference.[4] The interpretation of the TALE

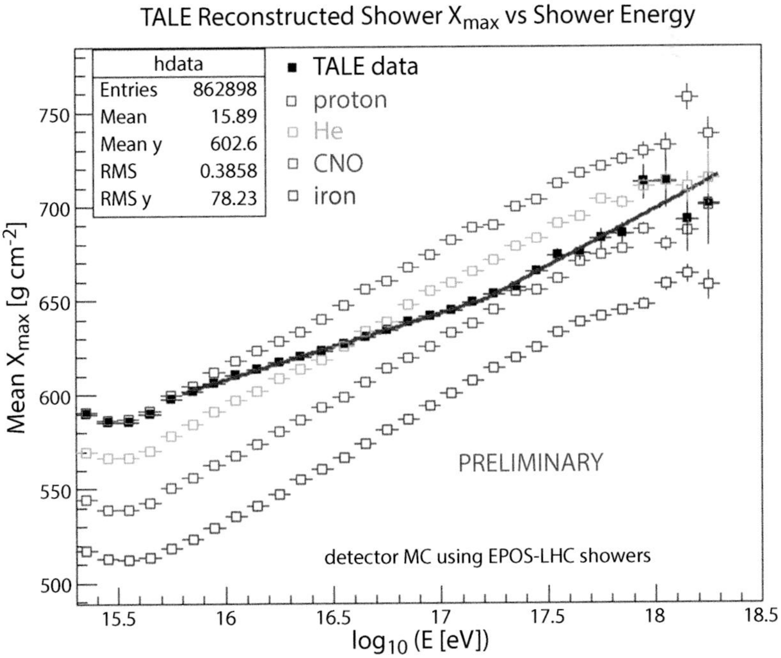

Figure 12.3   Mean X_{max} results from the Telescope Array Low Energy Extension (TALE) experiment, measured in monocular mode, as shown at ICRC 2019. This figure indicates that the composition is light near the knee, becomes heavier from about $10^{15.7}$ eV to $10^{17.2}$ eV, then at higher energies starts becoming lighter. From T. AbuZayyad et al., PoS(ICRC2019), 169; R.U. Abbasi et al., *Ap. J.*, 865, 2018, 74, with permission.

spectrum, described earlier in Chapter 8, is that the second knee at $10^{17.1}$ eV is taken to be the maximum energy of galactic accelerators for iron nuclei, and that above this energy galactic cosmic rays die away and extragalactic cosmic rays of light composition begin dominating the spectrum. This interpretation predicts that the composition should get heavier up to just above 10^{17} eV and then become lighter. The TALE $<X_{max}>$ results are compatible with this interpretation.

References

[1] R.U. Abbasi et al., *Ap. J.* **858**, 2018, 76.
[2] B. Peters, *Proceedings of the 6th ICRC,* Moscow, Russia, 1960, **vol. 3**, 157; A. Tomaszewski and J. Wdowczyk, *Proceedings of the 14th ICRC,* Munich, Germany, 1975, **vol. 8**, 2899.
[3] R.U. Abbasi et al., op cit.
[4] T. AbuZayyad et al., *PoS(ICRC2019)*, 169; R.U. Abbasi et al., *Ap. J.*, **865**, 2018, 74.

13

THE INELASTIC TOTAL CROSS SECTION

13.1 Introduction

This work has largely avoided discussing how the study of EAS can give information of interest to high energy physics. The reason is our emphasis on the astrophysical implications of the study of cosmic rays, plus the difficulty experienced in extracting reliable information on hadronic interactions from indirect measurements. Due to the low flux at ultrahigh energies, it is extremely difficult to study any but the most common processes. The proton–air inelastic total cross section $\sigma_{p\text{-}air\text{-}inel}$ is the most straightforward measurement one can perform. The relation between it and the quantity that is more interesting for high energy physics, the proton–proton total cross section $\sigma_{pp\text{-}tot}$, is model dependent. However, the energy dependence of the total cross section can be more reliably determined.

This particular high energy behavior is of interest to both high energy physics (in that it allows differentiation between various asymptotic models, as well as possible thresholds for new physics well beyond accelerator energies) and to cosmic ray physics, since the behavior of $\sigma_{p\text{-}air\text{-}inel}$ as a function of energy affects the position of X_{max} in the simulations of EAS showers and thus impacts composition measurements. The extraction of the cross section energy dependence from EAS studies leads to an important self-consistency test for EAS modeling. Since developing the concepts necessary for understanding this chapter *ab initio* would take us far afield, the reader with little background in high energy physics is invited to proceed to Chapter 14.

13.2 Relation between Inelastic and Total Cross Sections

Since EAS development depends on a significant exchange of energy between the primary proton and the air nucleus, cosmic ray measurements are not directly sensitive to the elastic, quasi-elastic, or diffraction dissociation components of the total proton–air cross section. The relation between these various cross sections can be written as[1] Equation (13.1):

$$\sigma_{p-air-inel} = \sigma_{p-air-tot} - \sigma_{p-air-el} - \sigma_{p-air-q.el} - \sigma_{p-air-diff} - \Delta\sigma \qquad (13.1)$$

where $\Delta\sigma$ is a screening correction that accounts for multiple scattering with excited nucleon intermediate states. The relation between the proton–air total cross section and the proton–proton total cross section is given by the Glauber multiple scattering formalism.[2] This formalism has been found to work well up to Large Hadron Collider (LHC) beam energies.[3] In the following, we assume that it continues to be adequate at PeV and greater energies.

Note that if the proton–proton total cross section is very large, then $\sigma_{p-air-inel}$ is not sensitive to it. This can be seen by ignoring, for the moment, the elastic and quasi-elastic corrections and using a simple Glauber picture in which the proton impact profile is neglected.[4] In that case, we obtain the expression as in shown in Equation (13.2):

$$\sigma_{p-air-inel} = \int d^2 b \left\{ 1 - \exp\left(-\sigma_{pp-tot} T(b) \right) \right\} \qquad (13.2)$$

where $T(b)$ is the impact profile of the air nucleus and b is the impact parameter. If σ_{pp-tot} is small, then the situation is shown in Equation (13.3):

$$\sigma_{p-air-inel} \approx \sigma_{pp-tot} \int d^2 b \, T(b) = A\sigma_{pp-tot} \qquad (13.3)$$

and we have a simple proportionality between $\sigma_{p-air-inel}$ and σ_{pp-tot}. However, if σ_{pp-tot} is large, the situation becomes as shown in Equation (13.4):

$$\sigma_{p-air-inel} = CA^{2/3} \qquad (13.4)$$

which is independent of σ_{pp-tot}! This is because the air nucleus becomes a black disk when σ_{pp-tot} becomes large, and the inelastic cross section becomes purely geometric.

A more precise calculation relating $\sigma_{pp\text{-}tot}$ to $\sigma_{p\text{-}air\text{-}inel}$ using Glauber theory requires the knowledge of a number of additional parameters.[5] These include B, the forward elastic scattering slope parameter; ρ, the ratio of the forward real and imaginary amplitudes; $\sigma_{pp\text{-}SD}$, the single diffractive cross section; $\sigma_{pp\text{-}DD}$, the double diffractive cross section, and $d^2\sigma/dtdM$ evaluated at t_{min} for the diffractive process $p + p \rightarrow p + X$ and lastly, the nuclear density profile. Despite the large number of parameters and the implied model dependence, measurement of the inelastic proton–air cross section can distinguish between rival models of asymptotic behavior of the total proton–proton cross section if the precision of the measurement is better than ~10%.

13.3 Measurement Techniques

We will discuss two techniques for measuring $\sigma_{p\text{-}air\text{-}inel}$: the X_{max} attenuation method (currently the most widely used approach), and the azimuthal angle dependence of fixed N_μ and N_e showers.

13.3.1 The X_{max} Attenuation Method

This technique assumes that the X_{max} position of a set of showers can be accurately reconstructed. For this reason, in the energy range of interest, it is usually restricted to air–fluorescence types of experiments. We saw, in Chapter 12, that the distribution of X_{max} in the atmosphere is sensitive to the primary cosmic ray composition. However, for $X_{max} > 800$ gm/cm^2 at energies > 0.1 EeV, we expect primarily proton induced showers. Given this sample of nearly pure proton induced EAS, we would like to measure the proton–air cross section by measuring the attenuation length of X_1, the point of first interaction of the proton. Unfortunately, since there is essentially zero light produced at this point, this is impossible to do. Instead, we can measure the attenuation length of the X_{max} distribution, which falls exponentially beyond about 800 gm/cm^2. One finds in Monte Carlo simulations[8] that λ_p, the proton attenuation length, is related to Λ_M, the X_{max} attenuation length, through simple proportionality as shown in Equation (13.5):

$$\lambda_p = K\Lambda_M \qquad (13.5)$$

where K varies from 1.15 to 1.2, depending on the hadronic model used. Once λ_p is known, the proton–air inelastic cross section is given by Equation (13.6)

$$\sigma_{p-air-inel}\left(mb\right)=2.4\times10^{4}/\lambda_{p}\left(gm/cm^{2}\right). \qquad (13.6)$$

The first attempt at such a measurement was from the Fly's Eye experiment at Dugway Proving Grounds, using both a high-statistics monocular data sample and a sample of data where tracks were viewed in stereo. Similar analyses have since been performed by the HiRes (stereo),[6] Auger[7] and TA[8] hybrid experiments. Since the position of X_{max} can be shifted by systematic reconstruction errors, care must be taken to use as unbiased a sample as possible. This is done for monocular data by applying tight cuts on the data and fitting the exponential slope above 800 gm/cm^2. Stereo data or hybrid data is analyzed in the same way, but with much looser geometric cuts due to the improved reconstruction. The resultant slope must be corrected for the effect of the acceptance as a function of X_{max}, but this is typically a small effect. The most recent results from the Auger and TA (see Figure 13.1) hybrid experiments are: Auger: 55.8 ± 2.3 gm/cm^2; TA: 50.5 ± 6.3 gm/cm^2.

Note that this result is not for a proton of any specific energy but, rather, applies to EAS with a spectrum of energies between the minimum and maximum energy cuts applied. Hence, to extract the appropriate value of the $\sigma_{p-air-inel}$ cross section from this data, we must assume an energy dependence.

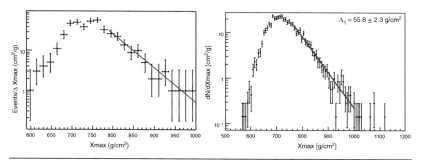

Figure 13.1   TA (left) and Auger (right) hybrid X_{max} distributions used to determine the proton–air cross section. From R.U. Abbasi et al., *Phys. Rev. D* 2015, 032007. (2015) and P. Abreu et al., *Phys. Rev. Lett.* 109, 2012, 062002., with permission.

Possible energy dependences of the cross section are shown in Figure 13.2. A Monte Carlo calculation can be performed using each assumed energy dependence to generate EAS. The calculation integrates over the cosmic ray spectrum from the minimum to maximum energy and accounts for the variation in the Fly's Eye acceptance with energy. The results for a simple hadronic model are shown in Figure 13.3. Each point in this figure corresponds to a different energy

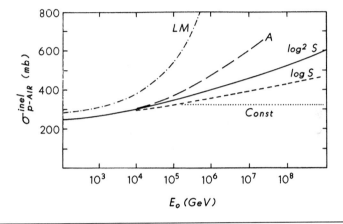

Figure 13.2   Possible asymptotic energy dependence of $\sigma_{p\text{-}air\text{-}inel}$.

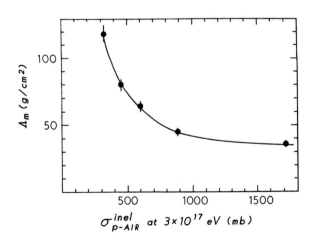

Figure 13.3   Relation between Λ_M and $\sigma_{p\text{-}air\text{-}inel}$.

dependence of the cross section which is arbitrarily evaluated at an energy of $s^{1/2} = 30$ TeV. Notice that as $\sigma_{p\text{-}air\text{-}inel}$ becomes large $- > 800$ millibarns (mb) – sensitivity of the slope of X_{max} to the cross section disappears. This is because the intrinsic shower fluctuations become larger than the slope produced by such a large value of the cross section. Sensitivity is good for $\sigma_{p\text{-}air\text{-}inel}$ between 300 and 600 mb, which is the most likely true range. However, different assumptions about the degree of scaling violation and hadronic model details change the precise nature of the curve.

For the case of the Auger experiment, the slope is compared to detailed hadronic model simulations but all hadronic model cross sections are multiplied by an energy dependent correction factor. The correction factor is then varied to find the best match between the predicted and measured Λ_M and the "true" value of the cross section determined.

The Fly's Eye and all the more current such results are subject to additional systematic effects. A large flux of alpha particles would be very difficult to distinguish from protons using the composition measurements techniques discussed in Chapter 12. If such a flux of alpha particles were present, the value of the cross section determined above would be an upper limit, since the actual proton component slope would be flatter than that which is measured. The use of different hadronic models such as QGSJet, Sybill and EPOS would also change the relation between Λ_M and λ_p. Experimental resolution will tend to make the X_{max} slope flatter and, hence, make the value of the cross section a lower limit. For current experiments, experimental resolution effects are small, however, since the stereo and hybrid data reconstruction errors are well-understood, and their effects can be well simulated.

More recent analyses by HiRes, Auger, and TA use more sophisticated hadronic simulation models, and also use different cuts on the data to minimize acceptance distortions in the slope of X_{max}. The energy range of the analysis is also restricted to energies where there is good agreement that there is a largely protonic composition. The current status of the measurements is shown in Figure 13.4

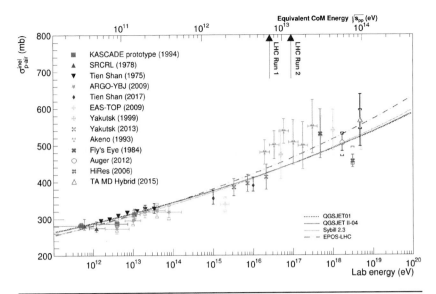

Figure 13.4   Compilation of proton–air cross section measurements. From R.U. Abbasi et al., *Phys. Rev. D*, 032007 92 (2015), with permission.

13.3.2 The Zenith Angle Technique

This technique was used by the Akeno array group to extract $\sigma_{p\text{-}air\text{-}inel}$.[9] The Akeno array measured N_μ, the total muon multiplicity, and N_e, the electron size at detection level. Since N_μ remains relatively constant as a function of depth beyond the shower maximum, its value at the surface can be used as a measure of the energy of the primary particle. On the other hand, N_e is attenuated beyond shower maximum. The electron size at observation level reflects the stage of development of the shower (i.e., the position of X_{max} above the observation level). Events with the same N_μ and N_e must have similar total energy and similar depth between first interaction and observation level (ignoring shower development fluctuations). The basic technique is to measure the attenuation of the number of showers with fixed N_e and N_μ as a function of zenith angle. This is given by the relation shown in Equation (13.7):

$$f(\theta) = f(0)\exp[-X_0(\sec\theta - 1)/\lambda_{obs}] \qquad (13.7)$$

where X_0 is the atmospheric depth at Akeno, $f(\theta)$ is the flux of EAS at zenith angle θ, and λ_{obs} is the effective attenuation length.

Since a mixed composition will distort this measurement, the experimenters introduce a bias towards deeply penetrating showers by demanding that for fixed N_μ, N_e be greater than some fixed N_{e_0} independent of zenith angle. They find that their result is not very sensitive to the choice of N_{e_0}. Once λ_{obs} is determined as a function of shower energy, it is corrected for the effect of shower fluctuations. This is done by generating Monte Carlo proton events and observing the dependence of λ_{inp}, the input attenuation length (representing the true proton attenuation length), and λ_{sim}, representing the measured length. The ratio $\lambda_{sim}/\lambda_{inp}$ is 1.5, and the Monte Carlo values of λ_{sim} are similar to the measured values λ_{obs}. The experimenters then divide λ_{obs} by the above ratio to extract the corrected value. This value is then used to calculate $\sigma_{p-air-inel}$ as a function of energy. Results are shown in Figure 13.4 (labeled Honda et al.).

There are clearly significant systematic effects present in the results presented by Fly's Eye, Akeno, HiRes, Auger, and TA. Even with the presence of these effects, the results are consistent with a *logs* to *log²s* growth of the cross section. The X_{max} attenuation method is the most reliable, and increased statistics in the future should give measurements of the slope as a function of energy.

13.4 Glauber Model Calculation

The relation between $\sigma_{p-air-inel}$ and $\sigma_{p-p-tot}$ is found through the Glauber formulation and was first examined in the context of cosmic rays by Gaisser et al.[10] They find:

$$\sigma_{p-air-inel} = \int d^2b \left[1 - \left(1 - \Gamma_A(b) \right)^2 \right]$$

where

$$\Gamma_A(b) = 1 - \left[1 - \int d^2b' dz \Gamma_N(b-b') \rho(z,b') \right]^A$$

and $\rho(z, b')$ is the density distribution of nucleons in the nucleus; and

$$\Gamma_N(b) = \sigma_{p-p-tot} \exp\left(-b^2/B \right)/4\pi B$$

where B is the forward elastic scattering slope. This relation must still be corrected for $\sigma^{q.el}$, σ^{diff} and $\Delta\sigma$. Their contributions can also be calculated. One can estimate $\sigma^{q.el}$ from the work of Glauber and Matthiae,[11] while $\Delta\sigma$ is a correction to the Glauber calculation itself, rather than to the data. This is due to the fact that the simple Glauber model used above sums over elastic nucleon–nucleon scattering amplitudes only and ignores higher order diagrams. These can be calculated to lowest order. The diffraction excitation cross section σ^{diff} can be related to the single diffraction scattering cross section σ^{SD}. Measurements of σ^{SD} and $\sigma_{p\text{-}p\text{-}inel}$ exist at up to LHC energies, see for example [12] They can be used to set a lower boundary on the value of σ^{diff}. Unitarity bounds, together with an extrapolation of σ^{el}/σ^{tot} from accelerator energies, can be used to establish an upper bound. The net result is that the diffraction dissociation cross section must be between about 10 and 20 mb at greater than 0.1 EeV energies. The total correction to $\sigma_{p\text{-}air\text{-}tot}$ ranges between 60 and 80 mb or a systematic correction of less than 20% for $\sigma_{p\text{-}air\text{-}tot}$ between 400 and 500 mb.

The remaining parameter for fixed nucleon density profile is B, the elastic slope. Figure 13.5 shows the relation between $\sigma_{p\text{-}air\text{-}inel}$ and $\sigma_{p\text{-}p\text{-}tot}$ as a function of B while Figure 13.6 shows the current data and expectations from hadronic models. Using the most recent data on B, Figure 13.7 shows the full set of results on $\sigma_{pp\text{-}tot}$

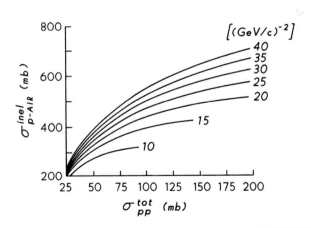

Figure 13.5 Relation between the p-air inelastic and total cross section and B, the elastic slope parameter.

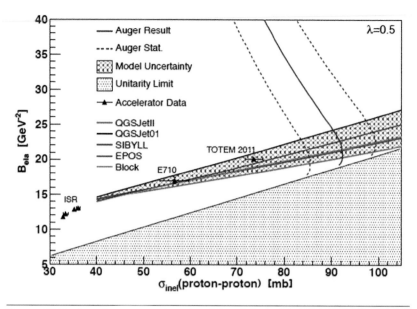

Figure 13.6 Current measurements and model predictions of $\sigma_{p-p-inel}$ and the elastic slope parameter B_{ela}. From R.Ulrich et al., EPJ Web of Conferences 53, 2013, 07005, with permission.

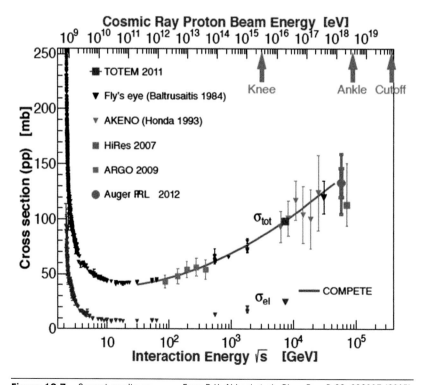

Figure 13.7 Current results on $\sigma_{p-p-tot}$. From R.U. Abbasi et al., Phys. Rev. D 92, 032007 (2015), with permission.

13.5 Discussion

Cosmic rays with energies near 0.1 EeV have similar center-of-mass energies as LHC proton–proton collisions. This means much more direct adjustment of hadronic model cross sections used for EAS simulations can be done at these energies. Current results using the X_{max} technique also give proton–air cross section values that are consistent with hadronic model extrapolations from accelerator energies. This gives more confidence that the extrapolations are also approximately valid for the composition studies too. In particular, the presence of an exponential tail in the X_{max} distribution consistent with expectations for protons is a strong constraint implying a significant protonic component in the cosmic ray flux in the energy range 1–10 EeV.

References

[1] T. K. Gaisser, U. P. Sukhatme, and G. B. Yodh, *Phys. Rev. D*, **36**, 1987, p. 1350.
[2] R. J. Glauber and G. Matthiae, *Nucl. Phys.*, **B21**, 1970, pp. 135–157.
[3] T. K. Gaisser, U. P. Sukhatme, and G. B. Yodh, op. cit., p. 1356.
[4] V. Barger et al., *Phys. Rev. Lett.*, **33**, 1974, p. 1051.
[5] T. K. Gaisser, U. P. Sukhatme, and G. B. Yodh, op. cit., p. 1350.
[6] K. Belov et al., *Nucl. Phys. B (Proc. Suppl.)* **151**, 2006, p. 197.
[7] P. Abreu et al, *Phys. Rev. Lett.* **109**, 2012, 062002.
[8] R.U. Abbasi et al., *Phys. Rev. D* **92**, 2015, 032007.
[9] T. Hara et al., *Phys. Rev. Lett.* **50**, 1983, p. 2058.
[10] T. K. Gaisser, U. P. Sukhatme, and G. B. Yodh, op. cit., p. 1354.
[11] R. J. Glauber and G. Matthiae, op. cit., p. 135.
[12] K. Goulianos, *Phys. Rep.*, **101**, 1983, p. 169.
[13] R. Ulrich et al., *EPJ Web of Conferences* **53**, 2013, 07005.

14

THE ATMOSPHERE

14.1 EAS Development in the Atmosphere

Since most of an extensive air shower (EAS) occurs in the atmosphere, it is important to understand the atmospheric properties that affect it. These fall into two categories: properties that affect the shower development, such as atmospheric density and composition; and properties that affect shower detection, such as atmospheric attenuation and scattering of light. We will begin with a consideration of the shower development issues.

14.1.1 The Troposphere

Although the actual atmosphere is a complex turbulent system, certain approximations allow a simple description that is generally adequate for our concerns.[1] The part of the atmosphere that is of greatest importance to EAS development is the **troposphere**, characterized by a decreasing temperature with altitude, extending from ground level to a height of about 15 km. The bottom of the troposphere often exhibits a homogeneous **boundary layer** that typically extends to 1–2 km and contains surface aerosols. At 15 km lies the **tropopause**, which is roughly 2 km thick and is where the temperature stops decreasing further, followed by the **stratosphere**, which extends out to about 100 km. The troposphere is characterized by turbulent behavior that generates the well-known phenomenon of weather. This complex phenomenon is produced by convective processes, with the heat source being the surface of the Earth. The temperature of the troposphere drops with altitude above the Earth's surface due to convective cooling, at a rate of ~6.5K degrees per kilometer (the temperature lapse rate). However, certain aspects of the troposphere can be adequately described by assuming that it is isothermal (the "isothermal atmosphere approximation"). In that case, it is straightforward to derive the relation of pressure to altitude.

14.1.2 The Isothermal Atmosphere Approximation

We assume that a small volume of gas of height dh and density ρ has a pressure change dP across it due to the gravitational force of the form:[2]

$$dP = -\rho g dh$$

The ideal gas law gives:

$$P = nkT$$

and

$$\rho = nm$$

so that:

$$dP/P = -(mg/kT)dh$$

and, finally,

$$P = P_0 \exp{-h/(kT/mg)}$$

where

$$H = kT/mg$$

is the scale height of the atmosphere. Although the atmosphere is not isothermal, the measured pressure variation in the troposphere follows an exponential form very closely. The scale height is usually assumed to be 7.5 km, although it can vary significantly from place to place.[3] The density profile of the troposphere follows a similar expression:

$$\rho(h) = \rho_0 \exp{-h/H}.$$

The density decreases by a factor of 10 for every 10 miles of increasing altitude.

The dependence of temperature on pressure, for the case of a constant temperature lapse rate, can be shown to be:

$$T \propto P^\alpha$$

where α is equal to $\gamma - 1/\gamma$ and γ is the ratio of the specific heats, which is 1.4 for the troposphere.

A more refined approximation to tropospheric density, pressure, and temperature can be found as the U.S. Standard Atmosphere.[5] If day-to-day and even hour-to-hour tropospheric variations are important, GDAS, an extensive interpolated data base is now available online.[6] For truly local and precise measurements, balloon radiosonde data is typically available at most airport locations.

The elemental composition of the troposphere is reasonably constant with altitude. The three most important components are N_2 (78%), O_2 (21%), and Ar (1%). There can also be significant concentrations of H_2O, as well as natural and manmade aerosols.

14.1.3 Atmospheric Slant Depth

Standard atmospheric pressure at sea level is defined to be 1013 millibars. This is equivalent to a vertical column of atmosphere of unit area weighing 1033 gm. Hence, sea level is said to be at an atmospheric depth of 1033 gm/cm². It is often important in cosmic ray physics to determine the atmospheric depth along a line inclined to the vertical. This is called the "atmospheric slant depth", and it is given by the expression:

$$X(\theta,h) = X_0 \exp(-h/H)\sec \theta$$

where X_0 is the vertical atmospheric depth at sea level, H is the atmospheric scale height, h is the height above which the slant depth is to be determined, and θ is the zenith angle of the line. This expression is quite accurate for zenith angles of less than 75°; at larger angles one needs to take into account the curvature of the Earth.

14.2 Atmospheric Absorption

The atmosphere affects optical EAS observation primarily by attenuating and scattering the produced Cherenkov or fluorescence light. Attenuation is due to both purely absorptive processes and scattering processes, which disperse light out of the direction of the line of sight. Absorptive processes are important at wavelengths of less than 290 nanometers (nm), where the ozone component of the atmosphere begins the strong absorption of ultraviolet, and above 800 nm, where water vapor and CO_2 absorption begin.[3] In between these ranges, the primary mechanism for attenuation is scattering by the atmospheric molecules themselves (Rayleigh scattering),[7] and by natural and man-made aerosol particles. Aerosol scattering is often approximated using a spherical drop approximation and is then referred to as Mie scattering.[8]

14.2.1 Rayleigh Scattering

Rayleigh scattering is the most straightforward of the two and we will discuss this first. The process is:

$$\gamma + \text{air molecule} = \gamma + \text{air molecule}$$

The cross section has a strong wavelength dependence:

$$d\sigma/d\Omega \propto 1/\lambda^4$$

The mean free path for scattering X_R at $\lambda = 400$ nanometers is 2970 gm/cm². We can write the number of photons scattered out of the beam per unit length as:[9]

$$dN\gamma/dl = -(\rho N\gamma/X_R)(400\,\text{nm}/\lambda)^4$$

where

$$\rho = \rho_0 \exp(-\lambda/H)$$

and

$$\rho_0 = 0.00129\,\text{gm/cm}^2$$

at 0K at sea level for an isothermal atmosphere.

The probability of scattering into a given solid angle has the following dependence:

$$d^2N\gamma/dld\Omega = (3/16\pi)[dN\gamma/dl](1+\cos^2\theta)$$

Rayleigh scattering has a rather slow angular dependence, making it important at all emission angles.

14.2.2 Aerosol Scattering

Aerosol scattering is quite complex. It has a rapid dependence on scattering angle, which varies with aerosol size, aerosol shape, and dielectric constant. The effect of this rapid dependence is exacerbated by the variable character of the aerosol content as it occurs in nature. The aerosol size distribution may change as a function of altitude, composition of pollutants, and weather conditions. Because of this complexity, no reliable predictions can be expected for atmospheric attenuation if the concentration of aerosols is large. For this reason, optical detection of EAS properties must be performed in regions where aerosol scattering remains a small proportion of Rayleigh scattering. These areas include the deserts of the western United States and Argentina, and high mountain elevations.

The theoretical framework for describing light scattering by spherical dielectric suspensions was formulated by Mie. A very much simplified model of aerosol attenuation applicable to desert areas is due to Eterman.[10] It is known from balloon flight studies that aerosols are concentrated near the surface of the Earth in the boundary layer. Within this layer, aerosol density is often independent of height. However, since the height can vary from day to day and hour to hour, integrated over time the density can be thought of as being distributed in an approximately exponential fashion with a scale height of 1.2 km. Although aerosol size distribution and shape are variable, an average aerosol model can be constructed that assumes a spherical aerosol size distribution such as:

$$N(a) = a^{-2.5}$$

where a is the radius of the sphere. In this case, one finds that the mean free path l_M is a strong function of wavelength and is approximately 14 km at $\lambda = 400$ nm. A useful expression for attenuation near this wavelength is, then:

$$dN\gamma/dl = N\gamma\left(\exp\left(-h/h_M\right)\right)/l_M$$

where h_M is the aerosol scale height (typically ≈ 1.2 km). The angular distribution of the scattered light is also a function of wavelength and aerosol model, but is strongly peaked at zero degrees emission and can be approximated by an exponential function:

$$d^2 N\gamma/dl d\Omega = a\left[dN\gamma/dl\right]\exp\left(-\theta/\theta_M\right)$$

where typically $a = 0.80$ and $\theta_M = 26.7°$.

Because of differences in their angular distribution, Mie scattering will dominate over Rayleigh scattering at small angles (for standard western desert atmospheres), while the reverse will be true for angles near and beyond 90 degrees.

The attenuation of light passing from a point at slant depth X_1 and height h_1 to a point at slant depth X_2 and height h_2 for Rayleigh and Mie scattering can be written as:

$$T_{Rayl} = \exp\left[-\left[X_1 - X_2\right]/X_R\right]\left(400\,\text{nm}/\lambda\right)^4\right]$$

$$T_{Mie} = \exp\left[\left(h_M/l_M\cos\theta\right)\left(\exp\left(-h_1/h_M\right) - \exp\left(-h_2/h_M\right)\right)\right]$$

and the total transmission is then:

$$T = T_R T_M$$

It is important to note that transmission coefficients for these processes can be multiplied only if multiple scattering of light is not important.[11] Situations in which aerosol concentrations are large cannot be adequately described with this formalism.

For days with good visibility (40 km) in the deserts of the western United States or in the Argentinian pampas, the contribution of Mie scattering is less than 20% of Rayleigh scattering.[12]

14.3 Measuring Atmospheric Transmission

Modern experiments using air fluorescence and Cherenkov light monitor variations in atmospheric transmission operate by placing calibrated "standard candles" at known distances from detectors. These include Xenon flash bulbs and YAG lasers producing light of known intensity and wavelength. Such light sources can be positioned at typical distances over which EAS are detected allowing the detector response to be monitored on a nightly or hourly basis. Such monitoring is also a valuable end-to-end calibration of the detector on nights when the atmosphere has very little aerosol and can thus be well-described by Rayleigh scattering.[13]

A stand-alone approach is to use a Lidar system.[14] Here, the detector is a separate telescope, typically aligned parallel to a laser beam, that measures the back-scattered light as a function of time. This provides a compact way of describing atmospheric scattering in a particular direction, though it does not include the cosmic ray detector response.

The presence of clouds in the observational volume of air fluorescence or Cherenkov detectors is problematic for the analysis of data. Clouds can affect the active atmospheric volume available for observation, as well as block or distort transmitted light. Various cloud monitors have been developed to give experimenters reliable hour-by-hour information on their presence.[15] These are typically infrared sensors mounted either as all-sky cameras, or as sensors positioned at each detector mirror. This information can be used to remove periods of data-taking where a clear and fully visible aperture is not available.

References

[1] R.C. Haymes, *Introduction to Space Science*, John Wiley, New York, 1971, pp. 54–90.

[2] R.C. Haymes, *Introduction to Space Science*, John Wiley, New York, 1971, p. 62.

[3] A.E. Cole, A. Court, and A.J. Kantor, in *Handbook of Geophysics and Space Environments*, Air Force Cambridge Research Labs, 1965, Chap. 2.

[5] U.S. Standard Atmosphere, 1976, *National Aeronautics and Space Administration*, Washington D.C.; http://www.pdas.com/atmos.html.

[6] https://www.ncdc.noaa.gov/data-access/model-data/model-datasets/global-data-assimilation-system-gdas.

[7] E.C. Flowers et al., *J. Appl Meteorology*, **8**, 1969, p. 955.

[8] L. Elterman and R.B. Toolin, in *Handbook of Geophysics and Space Environments*, Chap. 7.

[9] R.M. Baltrusaitis et al., *Nucl. Instr. Meth.*, **A240**, 1985, p. 414.

[10] L. Eltermanand R.-B. Toolin, op. cit.; L. Elterman, *Ultra Violet, Visible and IR Attenuation for Altitudes to 50 km*, Air Force Cambridge Research Labs, 1968.

[11] L. Elterman and R.B. Toolin, op cit.

[12] E.C. Flowers et al., op. cit., p. 955.

[13] Y. Takahashi et al., *AIP Conference Proceedings* **1367**, 2011, p. 157; The Pierre Auger Collaboration, *JINST 9*, 2013, p. 04009.

[14] T. Tomida et al., *NIM-A* **V654**, 2011, p. 653.

[15] J. Adam et al., *EPJ Web of Conferences* **144**, 2017, p. 01004.

Index

Page numbers in *italics* refer to content in *figures*.

153